Sister Bernadette
Cowboy Nun From Texas

To
Florence & Charlie
Smith

Love and prayers,

Sister Mary Bernadette Muller,
08

By Elizabeth Harper Neeld

Writing, 1, 2, 3 Editions

Writing Brief, 1, 2, 3 Editions

Readings for Writing

The Way A Writer Reads

Writing: A Short Course

*Options for the Teaching of English:
The Undergraduate Curriculum*

Either Way Will Hurt & Other Essays on English (ed.)

Harper & Row Studies in Language and Literature (ed.)

Fairy Tales of the Sea (ed.)

From the Plow to the Pulpit (ed.)

Yes! You Can Write (audio)

*Seven Choices: Taking The Steps to New Life
After Losing Someone You Love*

Sister Bernadette

Cowboy Nun From Texas

The Story of a Woman Challenged by God

Bernadette Muller
and
Elizabeth Harper Neeld

CENTERPOINT PRESS
Houston, Texas

Psalm Texts from *The Liturgy of the Hours*, translation, The Grail (England) 1963, Collins, London, 1963.

English translations of the Magnificat by the International Consultation on English Texts.

Design and Production by Dodson Publication Services, Austin, Texas

Front Cover Photograph by Sister Mary Angela Chandler

Back Cover Photographs, top to bottom, left to right, by Sister Mary Angela Chandler, Sister Mary Angela Chandler, A. Hussein, A. Hussein

Miniature horse on front cover photo and back cover photo (top right): BRW's Little Tonto; owner, Bobbie Ferraro

Miniature horse on back cover photo (top left): Monastery Celebration; owner, Jacqueline Gerland

Library of Congress Cataloging-in-Publication Data

Muller, Mary Bernadette, 1918-
    Sister Bernadette: cowboy nun from Texas / Mary Bernadette Muller and Elizabeth Harper Neeld.
      p.   cm.
    ISBN 0-937897-98-1 : $14.95
    1. Muller, Mary Bernadette, 1918- . 2. Poor Clares—Texas—Brenham—Biography. 3. Horse breeders—Texas—Brenham—Biography. 4. Miniature horses—Breeding. 5. Brenham (Tex.)—Biography.
I. Neeld, Elizabeth Harper, 1940- . II. Title.
BX4705.M82A3 1991
271'.97302—dc20
[B]   90-44791
                                        CIP

This story of my life is lovingly dedicated to those who gave me life: my father, George Henry Muller, and my mother, Nina Landis Muller....and to Mary Rita Kosch, my sister, and my brother, George Henry Muller II, who continue to bring joy into my life.

MARY BERNADETTE MULLER

Acknowledgments

Dorris Turner is responsible for the thorough research which informs each section of this book. Sister Mary Angela Chandler provided transcription of taped resource material as well as careful manuscript review. Cindy Spring, Ann Overton, and Lyn Fraser read the manuscript in its entirety and made insightful and valuable comments. Jerele Neeld provided media liaison services and design/production consultation. For the work of all these generous individuals, we are thankful.

Table of Contents

INTRODUCTION

By association, Texas and cowboys go together. But a **Nun Cowboy**? A contemplative life in the midst of a bustling horse farm, in the midst of running a thriving business?

At one of the recent American Miniature Horse Shows, Sister Mary Bernadette Muller received an unprecedented honor. Contemporaries in the miniature horse world gathered from all over the United States to name Sister Bernadette **The First Lady of the American Miniature Horse Industry**!

Testimonial letters acclaimed the "Cowboy Nun from Texas": her knowledge of miniature horses, her skill at operating a successful horse breeding and training ranch, her prowess as a business woman, her dedication and enthusiasm for the miniature horse industry. These same testimonial letters also acclaimed Sister Bernadette as a woman of God with values of prayer, sacrifice, humility, and poverty, as well as a woman whose joyful love of God shows itself in her every act and word.

Hence, what at first glance might look like a contradiction in terms—a sister named Bernadette and a woman doing cowboy work in Texas—are actually no contradiction at all. Sister Bernadette's work on the ranch is part of her work for her Lord.

All of us have our own story to tell; and when God is the moving force of that story, one can but exclaim, "How marvelous are Thy works, O Lord!" When Sister Bernadette re-entered my life three years ago—decades after she taught me piano lessons when I was a child, decades after she invited me to Bolivar, New York, to join the convent—little did I know the total story contained in the pages of this book.

Oh, I knew her mom when she came to Ft. Myers, Florida, each winter and I knew her sister who absolutely loved the gardenias that my papa let us pick to bring to school. I knew she had a brother and that she loved animals. In fact, there was always some creepy, crawly, jumpy creature who found a happy home in her classroom where, after school hours, for five years we spent many long sessions at the piano.

It was through her guidance and inspiration that I entered the Allegany Franciscan Community, but three years later her call from God to move into a contemplative community removed her from my physical presence for what I thought was forever. A letter which I received from her on my entrance into the Franciscan community contained a promise of life-long prayers for me that now, after nearly a lifetime and a re-acquaintance, I have come to value as a treasure beyond compare.

This special lady of God has a message for each of us. God's interaction in one's life has always been looked upon as a private affair, and to write personally and publicly about it can be interpreted as a lack of humility. In Sister Bernadette's case, this is far from the truth. When God's Spirit is alive and acting in our lives, there is a movement within the soul that leads us to speak in bold terms, to proclaim to all the world: "He has done great things for me, and holy is His name."

God chooses whom He wills, and when and where. Who are we to question His choice of bestowing the gift of contemplative graces on a Poor Clare nun in the 20th century, a nun who lives in Texas, who breeds and raises miniature horses to support the nuns under her care? She has proven herself to be a shrewd business woman, a competent spiritual leader, and an inspiring religious force in a world that hungers to know God's love.

SISTER KATHLEEN FRANCIS HONC, O. S. F.
LARGO, FLORIDA

Sister Bernadette
Chronology

Known by This Name	Dates	City	State
Nina Henriette Muller	1918-1934	Northfield	New Jersey
Sister Marie George (Muller)	1934-1936	Allegany	New York
Same	1936-1938	Rochester	New York
Same	1938-1940	Wood-Ridge	New Jersey
Same	1940-1946	Ft. Myers	Florida
Same	1946-1947	Bolivar	New York
Same	1947-1948	Haddonfield	New Jersey
Same	1948-1951	Niagara Falls	New York
Henrietta (Muller) then upon acceptance as Poor Clare Sister Mary Bernadette (Muller)	1951-1963	New Orleans	Louisiana
Sister Mary Bernadette (Muller)	1963-1985	Corpus Christi	Texas
Same	1985-present	Brenham	Texas

SECTION ONE

NEW ORLEANS, LOUISIANA
1951

Where and When My Life Took a Major Turn

CHAPTER ONE

FROM THE CONVENT TO THE MONASTERY

The law of the Lord is perfect,
it revives the soul.
The rule of the Lord is to be trusted
it gives wisdom to the simple.
FROM PSALM 19

Georgia
April 16, 1951
10:00 a.m. CST
. . . And Alabama

Dear Pop,
I love my pencil.
Train is taking a lot of curves so 'tis hard to write. The bed was so comfortable I stayed there until about 9 a.m. S. T. Got washed and dressed and then went to the car ahead for breakfast. Some couple sat at my table. Was glad. I had corn cakes — pancakes made with corn flour — maple syrup, 2 sausage patties, toast, jelly & butter, and 2 pots of tea. I made a mistake and poured syrup into one cup of tea instead of the cream from the cream pitcher. Prices sure are high on the train but I'm not caring. It is my last chance to splurge so I'm trying to enjoy myself.
I love my pencil.
Dogwood is blooming in profusion here. If I could have opened the train window I could have picked some wisteria a while back. The soil is redder than any I have ever seen. It's real clay. Yet the grass is green. Trees are green but quite a few are only covered with buds instead of leaves. Houses now have no cellars but are propped up on brick pillars. I've seen 3 television aerials!
I'm now eating orange life savers. I have my door closed, my linens off and my night cap on. . .very comfortable. We've stopped. 'Tis Lanette — not on the timetable nor on the map. However, I just saw a truck tag marked Alabama, so it might well be Alabama now.
We rode a little way and are now stopped again. This is Opelike, Alabama, still not on the timetable although it is on the map. All car licenses now have Alabama. I think that makes 14 states I've been in.
I love my pencil.

Now we're riding again. Red mud in abundance. Women outside wearing sweaters — some men have coats.

You fixed the hamster cage swell. Hope they'll be OK. Sure miss you all. Will be looking forward to October if that's when you'd rather come.

Will ask for same name for profession.

> Love,
> Nina

Alabama
1 hr. from Mobile
April 16, 1951
3:25 CST

Dear Sis,

Just pushed the bed back into place. Had a little catnap. Looks like the South now! Saw 2 big cabbage palm trees just a minute ago. The soil is so red and orange that I can hardly believe my eyes.

I love my pen but can't write straight on account of the train.

I'm so glad to have this roomette. It's perfect. I can control it at will. . . .

No more palm trees yet! Swamp, swamp, swamp; pines and more pines.

Sis, I sure enjoyed all the meals you prepared and all the other good things to eat. Bet you were glad to get back into your own bed. Last night lying in the berth I tried to estimate a width of 30 inches so I could imagine what it will be like on my new bed. Don't know what things I'll be able to tell you after I get there, but I'll do my best. . . . Just saw a huge "field" of wild flowers — deep maroon almost as large as tulips! I love my pen. Just saw a sign in swamp — "Sportsmen please don't shoot toward telephone wires." Wow — we're crossing a small body of water and I see big red colored ships (looks like warships)! Water must be deep but it doesn't look it. Spanish moss again. How they ever built a railroad through this swamp I'd never guess. Just saw first palmettoes and first blue crane or heron.

We're now just pulling out of Mobile, Alabama. I got off the train and walked to a mail box. Think I'll eat again before I get off the train. Hope you make out OK tomorrow, election day. I'll be anxious to hear. Gee, I miss you all. Hope you got home OK.

While I can tell you: phone me some day. I probably won't be able to ask you after I get there. But you could call and ask me to phone as

you've done before; then the superior would read the request. In that
way I could ask. Not often, though, on account of the $$.

Love!!
Sis

> You're finally here.
> Persevere. Persevere.
> You're finally here. . .
> Persevere...Persevere.

I prayed in rhythm with the rails as the train moved toward
the station. Within minutes I would be arriving at the place
where for so long I had been sure I wanted to be. But now my
mind swarmed with fears and doubts. I didn't have one single
idea of what life would be like beyond those walls. Would the com-
munity accept me? Would I be able to adjust to that kind of life?

I surprised myself with these anxious questions. For so many
years I had asked again and again for the opportunity to enter
the cloister walls. There had been the letters to the sisters in
Tennessee, to the sisters in New Jersey, and almost to the sisters
in Philadelphia—until I stopped by there one morning en route
home to visit my family in New Jersey and the sisters had served
me a raw fried egg for breakfast. I decided then and there that I
did not want to apply to any monastery that might serve raw
fried eggs every morning!

But in answer to all these earlier letters I had, on every
occasion, received a resounding *no.* When I wrote to the Poor
Clares at New Orleans, however, it had been different. Finally,
someone had supported my request. . . .

St. Elizabeth's Convent
Motherhouse and Novitiate
Sisters of the Third Order Regular of St. Francis
Allegany, New York

July 22, 1950

Rev. Mother Margaret Mary, O.S.C.
Monastery of St. Clare
New Orleans, La.

Dear Reverend Mother:
After prayer and deliberation, Very Reverend Father Juvenal
and myself have decided to give Sister Marie George an opportunity

to try the life of the Poor Clares if your Council should decide to accept her.

For over four years Sister has persevered in her desire and efforts to make this change. It would seem now that it is God's will. She is intelligent, has special talent in music and art, is refined, obedient, and seemingly very spiritual.

Sister has been refused admittance in the Bordentown and Memphis Poor Clares. I did not give too favorable a report to these places and this possibly was the reason. The Fathers and myself thought her to be a bit eccentric and over-determined to have her own way about this particular thing. However, since she has been so persevering in the face of many disappointments, we feel God may have special designs for her.

I send very kind regards and I ask God to bless your work for Him.

Yours very sincerely,
Mother Jean Marie
Mother General

In spite of the overall affirmative tone of the Mother's recommendation, I could still read the tentativeness, the uncertainty, in her words. I heard similar reservations in Father Juvenal's recommendation. . . .

St. Bonaventure College
St. Bonaventure, New York

Office of the President

July 22, 1950

Rev. Mother Margaret Mary, O.S.C.
Abbess
Monastery of St. Clare
720 Henry Clay Avenue
New Orleans 18, La.

Dear Reverend Mother:

This is in reply to your letter of July 14th, concerning the application of Sister Marie George O. S. F. for admittance to your monastery.

I have known Sister for the past three or four years, and as her Spiritual Director I am able to say that she is a very observant Religious. In view of this fact and the further fact that she has steadily persevered for these years in her desire to become a Poor Clare even in *the face of many obstacles and opposition, I offer the opinion that she merits at least the opportunity to test her desire and learn if it is a sign of a call from God to the contemplative life of the Poor Clares.*

May I ask you, Reverend Mother, and the Sisters to remember St. Bonaventure College and myself in your prayers,

Sincerely in St. Francis,
Juvenal Lalor, O. F. M.

Mother Margaret Mary in New Orleans had made it very clear that she, herself, wasn't certain this transfer was going to work out, for Mother Jean Marie had been required to promise that I would be welcome to return to St. Elizabeth's if things did not work out at the Poor Clare Monastery in New Orleans.

Yet in spite of the hesitancy expressed by my superiors, I had never, until tonight, had one moment's concern that this was the right thing for me to do and that I would manage well. But now that the train was pulling into the New Orleans station—now that a lifetime of being a cloistered contemplative nun was only a taxicab ride away—my heart was beating so hard I could see the movement of my habit as it rose and fell on my chest, rose and fell. My mouth was dry and the skin felt queer over my entire body.

My frame of mind wasn't helped at all when I was unable, upon leaving the train, to find anyone who would take me and my trunk to the monastery. "No, ma'am, can't do it. No way that big heavy trunk's gonna fit into this little cab." "Sorry, ma'am, wish I could but doctor told me not to lift . . . slipped discs in the back, you know." Finally, after being defeated again and again when I approached a driver to ask for help, I saw a cab pulling up to the curb to discharge a passenger.

"Well, I'll see what I can do," the driver offered. "I don't know if we can get it in the back or not, but maybe if we leave the lid up and tie the thing in with a rope. . . ." I breathed a sigh of relief when the kind driver and an equally kind porter finally managed to wedge the trunk into the back of the taxi and we pulled away from the station. As the cab began to move, I closed my eyes and leaned back against the worn leather seat of the taxi. The anxiety and doubt I had been feeling on the train had been replaced by a kind of quiet—I suppose you would say—anticipation and enthusiasm.

"Well, this will be the first and last time I'll ever see the streets of New Orleans," I confided to the driver as we drove through the city.

"You mean, when you go behind the walls of that monastery, they ain't never gonna let you out again?" the driver asked, incredulously.

"Well, it's not that they are never going to let me out again," I explained. "It's that I'm *choosing* to go in and never come back out."

"You mean, not even to go shopping or to go to church or to go see your family? Why in the world would somebody young like you do something like that?"

"In order to live a life of prayer. To devote oneself completely to God."

"But you're already a nun. Did you have to come all the way to New Orleans to devote yourself to God? Didn't they pray where you come from?"

I explained as simply as possible the difference between an active Order, such as the one I had been in, in which the sisters have vocations such as teaching, nursing, and social work, and a cloistered Order in which the sisters' main occupation is prayer. I told him specifically about my previous Order where I had spent my days teaching children music. I told him about the Poor Clare Order where the nuns take strict vows of poverty, chastity, and obedience, and commit to spend their entire lives in adoration of God, in supplication for their own souls and the souls of others, and in prayer for the will of God to be done in the world. "And that," I concluded, "is what I want to do for the rest of my life."

For a moment, I was afraid the man would ask me what had happened to make me want to change from teaching music to being a cloistered nun—and that was a story far too private, far too painful, a story I really didn't want to have to tell. But I need not have worried, for the cabbie's mind seemed riveted to the fact that this was the last ride in the outside world that I would ever take. *I was the last person that ever saw her outside those walls.* . . . I could imagine him telling his wife when he got home, talking in the hushed tones usually reserved for those conversations people have, after a death, about being the last person to see the deceased active and alive.

"You mean you won't go shopping or go back to see your family or go out to eat?" the driver repeated, evidently still trying to

understand the drama he inadvertently was taking part in.
"This is the very last time you'll see roads and cars and stores
and people on the street. . . . My, that's hard to realize," the
driver said. "That's hard to take."

"Well, my family will come to see me," I responded, reassur-
ingly, "and I'll be able to talk to them from behind the grill. And
I'm sure that if I ever were to get sick, really sick, they would
bring me out of the monastery to go to the hospital." (In fact,
during the next thirteen years the two times that I did leave the
enclosure were once to have my ears examined because they
were filled with noises and once to have an operation. That
operation, in fact, caused quite a stir. It seems that under the
influence of the anesthetic I grabbed the tie of the attending
physician when he came to check on me in the recovery room and
began to quiz him: "Know what this is good for?" I asked repeatedly as I jerked on the end of his tie. "To ho-o-o-ld onto your
man! That's what a tie is good for! To hold onto your man!"
Legend has it that the doctor left my bedside like a shot, not even
taking time to complete the examination. I do know that the rest
of the time I was in the hospital I never saw that doctor again!)

"Well," the driver said resolutely, "if this is the last time you're
ever going to see the streets of New Orleans, I'm going to take
the time to drive you around and point out some of the sights."

And before I could say yes or no, the tour began: Now, over
here's the ole muddy Mississippi. . . there's the St. Louis Cathe-
dral. . . this is Canal Street, where there's great shopping. . . and
there's Brennan's Restaurant where you can get the best food in
the world! The restaurant didn't look very impressive to me but
people were crowded all over the sidewalk waiting to get in, even
though it was almost midnight! So I agreed with the driver that
the food must be good.

I found everything interesting and quaint, and genuinely appreciated the driver's concern; but the truth was that getting to
the monastery was all that interested me. I was glad, therefore,
when the tour was over and we turned and began to drive toward
the monastery through the darkened residential streets.

"There it is," the driver called out. "There's the wall . . . that's
where you're going."

But I didn't think he was right. That wall didn't look like the
picture they had sent. "I don't think that's it," I told the driver.
Then a bit further down the street we came to the right wall. It
looked just the way it did in the pictures.

"I'm going to go around the block first," the cabbie told me. "That way you can see the whole thing before you go inside. Since you ain't ever going to be driving this way again."

―――――――

It didn't dawn on me until I was walking up the sidewalk toward the doors of the monastery that it would be very unlikely if anyone were to be in the front part of the building at midnight and that since my train had arrived almost three hours late no one would know when to expect me. So now the question was how was I going to be let in?

I began to knock. No answer. "Knock again, harder," I said to myself. "You've got to make somebody in there hear you." The taxi driver behind me was struggling with that big black trunk, twisting it first to this side then to that in a kind of rolling motion up the walk. "Come on," I implored, "somebody hear me. Somebody hear my knock."

Suddenly, the doors opened and there stood the three nuns whose job it was to interact with the outside world. "Oh, I hope you haven't been waiting here long. We're so sorry we didn't hear you," Sister Mary Roch apologized. "We were expecting you much earlier this evening, and when you didn't come we assumed you would be arriving on tomorrow's train. Oh, I would have felt so bad if we had not known you were here and you had sat outside until morning."

"Me, too," I chimed in. "But I was determined to knock until you heard me."

"But we didn't hear you . . ." one of the other sisters started to speak. Sister Mary Roch's stern glare interrupted her.

"Well, if you didn't hear me knock, how did you know I was here?" I asked, my curiosity getting the better of my discretion.

Glancing at the taxi driver who by now had twisted and pulled the trunk to the front door, Sister Mary Roch looked embarrassed. "Well," she said sheepishly, "you know how hard it is for the novices to keep all the rules . . . . One of the young girls was peeking out from behind the blinds in her upstairs room and saw you."

"I guess this is one time we can be thankful for someone breaking a rule," I replied to the sister, whose look told me my remark wasn't so cute as it was irreverent.

"Mother Margaret Mary is waiting behind the grill," Sister Mary Roch told me briskly. "She will tell you what you have to do." Then the Mother and I spoke briefly after which the three sisters joined me in a small room nearby where I was to change my clothes and make ready for the entrance.

———————————

From the big black trunk I removed the postulant dress Sister Jean Marie had given me before I left Allegany. Inside the small room I took off the habit of the Sisters of St. Francis and put on the garment which would indicate my desire and commitment to become a Poor Clare.

I felt undressed. After all, I had worn that habit for sixteen years! When I removed the tall, starched headpiece, I saw in the mirror a deep indention all the way across my forehead. A great big crease. "Maybe I'll have this dip in my forehead forever," I thought as I put on the old-fashioned-looking serge bonnet Sister Mary Roch provided me. I could feel the bristles of my sheared hair rubbing against the fabric of the bonnet as the sister tied it on me. Bristles, I would later learn, were a tipoff to the other sisters that I had come not from the outside world but from another Order.

When the nuns had finished assisting me with my dressing, Sister Mary Roch knocked on the inner doors. When the doors opened, Mother Margaret Mary was standing with a crucifix in her hand. I knelt on the floor before her.

"My child," the Reverend Mother said, softly, "You are crossing the threshold into God's own house for love. For love of your crucified Lord and Savior. If you always keep that one intention in all your thoughts and actions, you will receive the reward of that love, when you take your first step into that heavenly home and behold the beatific vision of your Spouse and Savior." These words sounded very beautiful to me and made an intensely strong impression. I kissed the crucifix in Mother Margaret Mary's hand and then stepped across the threshold into the enclosure.

In the long hallway of the cloister, waiting, stood a whole row of little old nuns—well, to be accurate I suppose I would have to say they were not all that *little*, for there were all sizes and shapes—but most of them were old. The sisters had just finished saying the Divine Office for their midnight prayers, and the younger nuns had been sent up to bed. But here were the professed sisters all lined up, waiting for me to enter.

At that moment, I remembered the anxious thoughts that had plagued me as the train pulled into New Orleans. *Had I done a foolish thing? Would I be able to make it?* Looking at these nuns now, however, I knew I would do everything within my power to get to stay here. "Look at them," I thought to myself. "I think I have never seen such happiness. Such calmness. Such peace. These are old sisters. And when I am old, this is the way I want to be. This lets me know I am in the right place."

Then Mother Margaret Mary led me to the end of the hall. One by one, I passed in front of the professed sisters. Right down the line I went, each sister greeting me warmly with the Franciscan hug. "Let us go to the chapel now," the Mother spoke quietly to all of us when I had reached the last sister. We all filed into the chapel behind the Reverend Mother. Kneeling in front of the altar, the lit candles throwing a bath of soft light on the beautiful tabernacle and the statue of the Virgin Mary, we prayed the prayer of the Holy Blessed Mother:

> My soul proclaims the greatness of the Lord,
> My spirit rejoices in God my Savior
> For He has looked with favor on his lowly servant.
>
> From this day all generations will call me blessed:
> The Almighty has done great things for me,
> And holy is His name . . .

I didn't think, kneeling there, that I had ever lived a more beautiful moment.

CHAPTER TWO

THOUGHTS AND ACTIVITIES OF A NEWLY CLOISTERED NUN

> Walk through Zion, walk all around it . . .
> examine its castles,
> that you may tell the next generation
> that such is our God,
> our God for ever and always.
> It is he who leads us.
> FROM PSALM 48

April 19, 1951

Dear Papa, Sis, and Bruds,

This is my first letter to you from the cloister. I have just finished reading my first letter received, from Sis.

To tell you the thing you may first want to know — this is where I belong. I've never felt so sure of anything in my life. Of course, I know the future is in God's hands. With His grace I expect Final Perseverance as a Poor Clare.

On the night I arrived Rev. Mother spoke to me in the parlor first to explain the procedure and told me I shall be called Sister Henrietta until reception (at least six months). Not until I was standing outside the door, waiting for her to open it, did I realize my second baptismal name was Henrietta. Not having used it, it still sounds strange. Aunt Henrietta would be pleased, I guess, and so should you, Pop — but I still like George better.

I knelt at the threshold while Rev. Mother Margaret Mary handed me a large crucifix to hold while she said an entrance prayer and welcome. I then kissed the crucifix and stepped inside my final home to meet some of the sisters and my Mother Mistress. We then proceeded to chapel for a few short prayers, then to the chapter room where I knelt before Rev. Mother who placed on me a cord and the Franciscan crown (rosary).

Then my Mistress showed me to the room I would occupy. The bed is at least 42 inches wide instead of the expected 30! There is a nice combination washstand with a drawer. I am using it as a desk at present. The chair has a little shelf under the seat. This I discovered recently. Clothing hangs on five hooks behind the door. There is a window almost reaching from the ceiling to the floor and the ceiling is extremely high. I think my room must be located the last one on the extreme right in front of the house as marked in the picture. I can't be certain, though, for the front of the house outside is not included in the enclosure. We go out on the grounds by going out the back door. (I'll tell you about the grounds in the next letter. Although once a month is the usual regulation for writing a letter, Mother Mistress has given me permission to write to you soon again to reassure you all is well and I love my new home.)

Needless to say I slept very little (about 2 hours) the first night. Ever since I've been like Rip V. Winkle at night.

The other morning I came into my room to find Mother Mistress climbing down a ladder. She had replaced a crucifix with the big one I brought. Every sister has been so kind. I have never experienced such genuine charity and courtesy. The monastery is a real haven of peace. I can never be grateful enough to God for having led me here. It was well worth the waiting.

Don't for one minute worry about me losing weight! The lack of meat could never be noticed on account of the abundance of vegetables and soups very carefully prepared. My first dinner (I could still see that plate when I had finished collecting) consisted of 2 fried eggs, string beans, white potato, rice, a whole tomato, corn & peas, bread — preceded by a bowl of vegetable soup — tea, water. I'll surely develop rosy cheeks!

One of the most beautiful surprises was the first morning (and every morning) when the chapel was prepared as for benediction. Rev. Mother approached the altar, incensed the Blessed Sacrament and opened the tabernacle doors back. There was a beautiful monstrance, already containing the Sacred Host. We sang O Salutaris, prayed, had a half hour meditation, said the rosary and sang Tantum Ergo before Rev. Mother closed the doors — without, of course, touching the monstrance.

I could not possibly tell you all in one letter.

It has taken me three sittings to write this.

I had the privilege of entering in time for a retreat! At the end of which (Saturday) there will be a Reception and Profession. I have the privilege of playing the organ (manual and electric) for the Mass and Profession since the organists are being professed. Retreat started today, but I had permission to finish this letter so you wouldn't be worrying about me.

The pen writes beautifully.

The trunk traveled in good shape and its contents were much appreciated. Father McCallion's gift — the Blessed Virgin Mary's statue — is in the Novitiate. Tell him.

Rev. Mother says the projector will be a treat and told the sisters at table that we would see it after retreat. Mother Mistress has never seen anything like it; she was already intrigued with the hand machine. I'm sure the projector will delight these dear Sisters.

Spanish Moss hangs in the garden — but that will have to wait.

Must hurry now. I love it here. I belong here. I know it. Of course, I miss you all but as I said before, I can help you more here than I could at home in New Jersey!

Pray that I shall be granted final perseverance and be granted necessary graces to overcome little difficulties which would be expected in this life as in any others.

Love and prayers,
Nina

Address envelope — Sister Henrietta.

Overcome little difficulties. Now that was going to be the challenge.

In the most unexpected ways and often at the most unexpected times, I found myself breaking some rule, acting out of line, or inserting some foolishness into the serious life of the monastery. Not many days after I had arrived, Mother Mistress called me to join the other postulants for night prayer in the novitiate. It had seemed to me in the days since I had arrived here that we never got through praying. There were the hours of prayer at daybreak, midmorning, noon, midafternoon, vespers, and midnight and, in addition, extra community prayers. It seemed that every Reverend Mother who had ever served this monastery had had her own favorite prayers which she added to the daily list—never, of course, bothering to remove any of the previous Mothers' favorites! I enjoyed praying, but kneeling hour after hour on the hard wooden floor with no cushion underneath . . . I just hadn't gotten used to it.

"May I be excused to go to the bathroom before we pray?" I asked Sister Helena Clare when she called us this particular night to the novitiate to pray. "Okay, go ahead," she said. "We'll just kneel here and wait." So I hurried; and when I came back, I knelt down and I guess I had a funny streak in me. I said, "Do you know what I was wondering?" Sister Helena Clare asked, "What?" imagining, I suppose, that I was about to ask some significant religious question. "I just had to put a new roll of toilet paper out because the old one was finished," I responded, "and on the wrapper it says, 'One thousand sheets to a roll.' I wonder if anybody ever really counted to see if they are being gypped." With that, I started laughing uncontrollably, as did the other sisters. We were all now so tickled that we were falling over on each other, laughing. It was minutes before the Mother Mistress could quiet herself and the rest of us so that we could carry on with our praying.

Then there was the matter of the bow on the bonnet.

All of us postulants had to wear our little old-fashioned-looking bonnets at all times when we were out of our rooms. (Even in our rooms we wore a night cap to keep our head covered and a towel over our head if we were going to the shower—a Poor Clare's hair could never be seen or her head uncovered.) Well, this postulant's cap gave me a lot of grief. The cap had little fluting all around the face, and under the chin there were two long streamers which you had to tie into a bow every time you put the cap on. The mistress of the novices inspected that bow because it had to be just so—no wrinkles in the streamers whatsoever, bow tied perfectly straight.

One day I decided that I had had enough of this. Slipping down to the laundry room, I surreptitiously ironed the streamers, hoping against hope that no one would see me, and then hurried back to my room where I did an alteration job. "I'll fix this aggravating thing once and for all," I said to myself. So, first, I tied the freshly pressed streamers into a perfect bow. Then I cut the bow off on one side where it was attached to the cap. I put a snap on that side and also a snap on the cap. Then all I had to do when I took that cap off was to unsnap it and the bow came loose on one side and stayed together in perfect shape.

Well, I guess I had worn my cap this way for several weeks when the mistress of the novices stopped me one day. "Your bow always looks so neat," she said, suspiciously, "and I know you haven't asked me permission to press it. How do you keep it that way?" "Oh," I said, "I had a good idea—look what I did." Then I proceeded to unsnap the cap and show her how I had fixed a permanent bow. "You know you shouldn't have done that," Sister Helena Clare scolded me. "I should make you take it apart and sew it back the way it was intended." But, if the truth were known, I think she thought it was a good idea because the only other thing she said was, "Go on now and be about your business."

On another day, I was sure I had gone too far. The monastery had the poorest excuse for an elevator that I had ever seen in my entire life—a rickety contraption with very little sides on it, barely able to hold two people and then only if they were skinny. It was a scary ride to take that thing from the basement up to the attic.

But this is what we had to do on wash day if it were raining because when the clothes couldn't be hung outside we had to hang them on a line strung in the attic. Well, one unlucky nun would be selected to ride up in that rickety elevator with the basket of heavy wet clothes beside her—then everyone else would make a dash to the top floor and get hold of the rope up there and pull it so the elevator would come from the basement to the attic. I'm telling you it wasn't easy.

Well, one rainy day the mistress of the novices remained down in the basement to finish up the wash while we novices went up to the classroom for a religious class given by Sister Mary Michael. Maybe it was because it was a rainy day, but we all seemed to be awfully weary sitting there listening to the sister drone on. The desks were old fashioned, facing one another in a row, and had a top which we could lift up to retrieve anything we might have stored there—books, our sewing or handwork projects. This is what we did during "recreation"—that was the big "recreation deal"—to do some work with our hands.

But this was a regular class now, and we were just all sitting there listening. Suddenly I thought to myself, "While I'm listening to her drone on, I could be doing something worthwhile with my hands." So I raised my hand and said, "Sister, may I please open my desk and get out my tatting?" I had recently taught myself to do some tatting. "Why, what is the matter?" she responded. "Well, Sister, I'm bored." I could hardly believe the words had come out of my mouth—but there they were. Sister retorted, "What, you are bored with God?" "No, Sister," I answered. "I am not bored with God. I am bored with you." At that, all the other novices gasped—I could hear them take in their breath in shock at what I had said. And at that exact moment I looked up. Standing in the doorway was the mistress of novices and, believe it or not, she had a twinkle in her eye.

"Would you all please come here," she said very quickly. "I'm ready to hang the clothes in the attic." Well, we all breathed a sigh of relief and got up and left. When I was the one told to get on and ride that old elevator up with basket of wet clothes, I didn't complain at all. That was one day I was glad to do it. I don't know what would have happened if the mistress of novices hadn't showed up when she did. But nobody ever said anything to me about my remark, although they certainly could have!

———————

Thoughts of my family, of course, were always with me. I had placed in my missal the letter Papa had written me in the days immediately after he, Sis, and Bruds put me on the train in Philadelphia. I could not count how many times I had read it.

Monday, April 16, 1951

Dear Nina,

We arrived home on Sunday at 20 minutes to seven o'clock, passing through several rainclouds. However, it had not been raining in Northfield.

This morning, on my way to work, I observed the clover which is spreading profusely over our lawn, and I recalled the day when you commented upon the rapid spread of the clover. The thought came to me this morning that this clover is symbolic of the many years that you have been a member of the Third Order of St. Francis. As the tiny clover seeds that fell upon the fertile soil of the lawn grew and developed and, in turn, scattered their seeds until the entire lawn became covered, so, too, have your years of devoted service in the convent (teaching the children in the articles of our Catholic Faith) borne fruit. The seeds of Faith which you sowed in the classroom, your example, found a fertile soil in the hearts and souls of the children and as the years pass, these children, grown into manhood and womanhood and many of them bringing into this life more children, the seeds of your teaching will spread and continue to spread. Nina, yours has been a most fruitful life as a member of the Third Order, and the most ardent wish and hope of your father is that you will find joy and happiness in the new life you are now entering.

A woman once approached Abraham Lincoln and told him that she had several sons in the Union Army, and said that it was her hope that God was with the cause of the Union. Lincoln meditated for a while and then replied that he was not so much concerned as to whether God was with the Union cause, as he was if the Union cause was with God. If, in God's wisdom, this chosen field is God's will, then you will find joy and happiness in your new life. Otherwise, you will not be contented and this discontent will be God's way of expressing His wish. He will then convey to you that it is His desire for you to devote your life work among His children, in teaching and by your example lead them to the preservation of their Catholic Faith.

It is your hope that God is calling you into this new life, but it is for Him to decide if you are to live a contemplative life or a life of teaching. You have made a noble decision that, in itself, surely has God's blessing, but if, at some future time, you become convinced that God is calling you for the life you have had during the past years, you should respond to His call, and realize that your true vocation is in the Third Order.

Child, I make no requests of you, but one I do make: that if the time comes when you do not experience the joy and happiness which should come to you if it is God's wish for you to lead the new life, you will return to your former vocation, happy in the thought that you did strive to serve God in the contemplative life. I want you to promise me just that one thing — that you will return if you feel in your heart that God chose your former life for you in serving Him. Remember, our Blessed Lord said that suffer the little children to come unto me, and what you do for them you do for Me.

Nina, you need not worry about Mary. She shall never want as long as I live, and should not want after I have gone.

I hope you had everything you desired during your short visit home. I shall miss going with you to Church in the mornings. Do you know that I did not go with you only one day, and that was on the first Saturday of April when Mary went with you. Of course, I did not receive Holy Communion on the first day because it was the day after your arrival home and we thought you would go to a later Mass. So I had something to eat that morning just before you came out of your room and said you would go to the 7 o'clock Mass. I am glad that I could go with you every day that you spent home on your vacation and hope that you will remember me in your morning prayers. I say the Rosary every morning in memory of your Mother, and on every Christmas I make a donation to the Church in her name and my name.

****

These stars represent a lapse of one day, yesterday, Primary Day. At 6:15 o'clock in the morning I had the police call for me to inspect our new voting district in the Fire Company garage. Bud Lever helped me to set up the booths and place the ballot boxes and arrange everything for the opening of the polls at seven o'clock.

Mary and I had dinner at one o'clock at home. I made one trip to Linwood for correction of a district in the afternoon, visited the polling places, and about 4:00 returned home. Bruds came home at about 5:30 and he had his dinner — crab cake and vegetables. The members of the Board reported to me in Pleasantville where I received all the books of the municipalities along the Shore Road. I got home about 12:40 a.m. this morning.

Yesterday Mary received your letter and we enjoyed the folder which explained in some detail the advantages of the roomette and which you had somewhat improved (?) with your artistic embellishments. We are awaiting subsequent reports of your journey. Yesterday I mailed (airmail) a letter to you from Mary.

This brings the news to the present moment, just 2:20 o'clock p.m., Wednesday afternoon.

You will hear from me more frequently and I and all of us hope you will be able to write to us at frequent intervals.

With love,
Your father

P.S. If there be anything you want for which you may ask, just let me know.

I take great consolation in knowing that you offer up your prayers and Masses for your mother. I hope that you will not forget me when my time comes. I need your prayers even more than does your mother. There are not so many mothers and wives today who are what your mother was. She was God's greatest earthly gift to me, so unworthy of her. I say my Rosary daily for her, every morning either at home or in Church, where I now go every morning during the summer months. . . .

(Papa)

I recognized the deep pain in my father's letter, his concern for my well being, and even perhaps his doubt that I had done the right thing in leaving the noble profession of teaching young people in order to join the Poor Clares. Papa's pain, in turn, pained me. But I knew I was in exactly the place God wanted me to be—although I had no earthly idea why He had sent me to the cloistered life. I wrote as often as Reverend Mother allowed in those first few weeks. . . .

April 29, 1951

Dear Pop, Sis, Bruds,

Have received all your letters. First of all — to answer your questions before taking you out into our cloister garden. Papa, you may be sure I would live as my conscience directs. Right now I can say in all sincerity — this is the life for me — it is what God has ordained for me from all eternity. I shall be a postulant until October — then one year as a Novice — then I shall take Solemn vows instead of Simple vows since I am already with Perpetual vows.

Certainly by the time Solemn vows are to be taken my decision should be well grounded and the vocation tried. But I'll have you know I feel perfectly at peace — for the first time in my life! And God grant that each day I receive sufficient grace to do His Holy Will as a Poor Clare.

Bruds, so glad you are learning teller work. Stick to it now. I'm praying for your material success every day. The name of the book you requested: The Graces of Interior Prayer by A. Poulain. S. J. B. Herder Bk. Co. 17 & 17 South Broadway St. Louis, Mo.

Sis, Reverend Mother says if it is not too much of an expense we'd like very much to have that screen. Today I shall prepare some of the reels so they can be shown in some sequence. In my next letter — probably not for a while — I shall be able to tell you how the nuns enjoyed the pictures. I'm sure they will be delighted.

You asked the phone no. It is not listed but is Uptown (Exchange) 3160.

There is absolutely nothing I need now. Mother Mistress says that if anything presents itself we'll let you know.

Now — come with me for a walk in our monastery garden. We go out the back door (either one). Someone comes bounding up to greet us. That is Boots, the big collie dog. She is very friendly and a beautiful creature. Now, look up! Towering live oaks border the outer wall on one side and Spanish moss hangs over the wall into the garden. These trees are twice as large as the one behind the church in Ft. Myers. We'll walk up one of the cement walks. Thick clover and grass make a lovely lawn over the whole garden. Rose (beds), carnations and violets, some glads, huge wild poppies (as big as large tulips) and oleander are now in bloom. There are three palm trees and I believe some of the fruit trees are orange and lemon. There is a small vegetable garden, a large chicken yard (about 30 hens and 1 rooster) and duck yard (about 20 ducks).

There is also a crypt. (I think that's what you call it.) A little "house" where the nuns are buried in vaults above the ground. It is very nice. I think there are about half a dozen there. The grounds are well cared for. One of the sisters looks after them. Ben is gardener and "handyman." There is a greenhouse — all glass . . . I just went down to pick you a violet and discovered there are many, many glads not blooming yet and several large beds of yellow daisies in a profusion of blooms. One of Boots' favorite pastimes is chasing the pigeons which make their home here. Some are coming right under my window now.

I was most fortunate to have arrived in time for the retreat. It was given by a Franciscan from Chicago. Friday we had a reception. Now I am the only Postulant & Saturday I was privileged to play for the ceremonies — a solemn profession during a Solemn High Mass (3 priests) followed by the ceremonies of a Simple profession and Benediction. Both sisters received a crown of thorns, the former wears hers for three days. After Compline last night we led her in procession to her room, singing the Magnificat. It was beautiful. I shall be the next, please God. Rev. Mother said I could make my own crown of thorns if I grew the bush, so, Sis, please ask Helen to send me one, as big as possible so I can start growing it. When she prunes the big bush maybe you could also send me a number of branches to make crowns for the future. I think they would still have to be fresh by the time they arrive here for they crack when you bend them if they're dry. In any event, ship a bush to plant. She promised me one.

Sis, will you please drop a line to Sr. Conrad, St. Elizabeth Convent, and Margaret, Sr. Kathleen Francis, Holy Cross Convent. I'll write to them at Christmas but you could repeat the contents of my letters. I know they are anxious to know how happy I am. Tell Sr. Conrad thanks for the sweet nuts and holy cards. I prayed for her on her feast day. If you have time, you could write to Mother Vincent Marie, St. Mary's Convent. She was always my ideal and inspiration and I was her postulant 17 years ago! Remember, I had her in the 8th grade.

So the house is painted! Let me know how the bookcase looks if you really have it put in. There are some books on music I left in those last cartons — donation — ha!

This pen is a wonder and the pencil has been used all week during retreat!

Keep praying for me. I couldn't be any more content.

Glad the pansies are still blooming. I enjoyed planting them on the grave.

Don't worry if you don't hear from me for awhile. I'll write again when I can.

Love and prayers,
Nina

SECTION TWO

NORTHFIELD AND PLEASANTVILLE, NEW JERSEY
1925-1934

Life As A School Girl

CHAPTER THREE

AN INTIMATE PEEP INTO MY EARLY CHILD-HOOD AND MAKINGS OF A VOCATION

> I think of your name in the night-time
> and I keep your law.
> This has been my blessing,
> the keeping of your precepts.
>
> FROM PSALM 119

I must have been in the third grade when the talk first got started about a Catholic school's being built in connection with St. Peter's church in Pleasantville. I know I was in the third grade when I started worrying about it. "But, Mama," I would say, "I love Mrs. Boddy . . . she's the best teacher in the whole world. . . . I don't ever want to be in anybody's room except Mrs. Boddy's."

In public school we could do things unheard of at home—like drink milk through a straw out of a cardboard carton instead of using a glass! We could play the instruments in the rhythm band as vigorously as we wanted to—the sounds weren't anything like the music at home—and I was one of the loudest parts of it. When we sang "Mine eyes have seen the glory of the coming of the Lord," the brick foundation shook, and I have no doubt the Lord and all Northfield heard. And Papa was not there to frown on my boisterousness! I also felt a sense of freedom from the watchful eyes of Mama as I stood—yes, stood—precariously on a chair with a huge set of cymbals on high. Yes, home had been fun; but public school was a new kind of fun.

I had not started school until the age of seven, since my parents believed the early years should be spent at home. But I had used pencils, crayons, and Mama's primary reader from very early, and I could not remember not being able to read or recognize numbers.

School, however, afforded me opportunities to share secret abilities that, up until then, had been bottled up inside me. How glorious the day when my minute observation of the gelding pony we had at home paid off! All across the blackboard border in our second grade classroom paraded cutouts of my drawings of ponies! These drawings of mine which the teacher had selected for display made the classroom beautiful to my eyes.

Competition was a game. How amazed I was when a third grade classmate stood up and offered to spell "Mississippi"! And how surprised I was when for a free-choice spelling word I chose "answer" and got it wrong because I left out the "w." The first incident was a means of encouraging me to tackle many scholastic problems which I might have chalked up as impossible; the second incident made me cautious to check facts before being so cocksure of myself.

My teachers had all been gentle women. The first I liked because she recognized my love of drawing and animals and she had such a sweet disposition. The second had been so cheerful and happy that she made school a delight. And now in the third grade my love for reading and poetry was growing. Excitement mounted when, reading books in hand, we formed little groups in circles and in subdued voices went round and round until we unwound the story and ourselves. "Reading and writing and 'rithmetic, taught to the tune of the hickory stick" . . . I loved to sing that song. But when I found out what a hickory stick was used for, I could never figure out why it would be needed. Learning was fun!

But now I was going to have to leave my little dark-haired best friend, Marion Breunig. (In truth, she was to remain a lifelong friend.) I was going to have to leave Mrs. Boddy who made poems and stories so much fun and leave these happy classrooms that were so familiar.

I was going to have to go to another school.

"But," Mama reassured me, "don't worry ahead of time. You won't be transferring until you're in the fourth grade . . . and, besides that, you will love the new school . . . there you will have sisters who will be your teachers!" Then she would tell me more: The sisters would be wearing black veils but they were brown sisters . . . they were coming from other places to live at the school and teach us . . . we would study our regular subjects like English and science but we would also study the lives of the saints and the catechism . . . oh, I would love it so much, she assured me.

Mama didn't know it, but the thing that piqued my interest the most was when she told me we were going to be taught by the brown sisters. Imagining as only an eight year old can, I began eagerly to await the momentous occasion when we would get to meet these teachers. Finally, the announcement arrived. "You are invited," the card read, "to meet the teachers of the new St. Peter's School. We are eager for your presence."

All of us, dressed in our finest, filed into the formal Governor Smith Room at the school to pass along in front of the sisters—Sister Vincent Marie among them—shaking each teacher's hand. I was surprised and shocked. As soon as we had shaken hands with the last sister—but long before we were out of earshot of Father McCallion and the teachers—I took issue with Mama: "You told me that they would be brown sisters. But everyone of them is white!" Mama and Papa, embarrassed I am sure beyond words, shooed me out of the room as quickly as possible all the while that Mama tried to quiet me by explaining. "Brown refers to the color of the sisters' Franciscan habit, not to the color of their skin!"

My first day in class at St. Peter's did not begin much better. "Turn in your geography books to the map of the United States," our teacher instructed, as she handed out drawing paper and pencils. "When you have found the map, draw it free hand on this paper and when you've completed the outline, fill in the states." I took immediately to the task; I would do this so well the sister would be bound to like it. But was I ever to be surprised.

As we proceeded, the teacher walked up and down the aisles to observe our work. When she got to my desk, she scrutinized my map for several seconds. I sat, proud of my work and waiting for her praise. Instead, with tight, pursed lips, the teacher picked up my map and placed it on top of the outline of the map in the geography book. She thought I had traced it! "Why, she doubts that I can draw that well," I thought to myself. I was so hurt. It felt terrible not to be acknowledged.

I was not unhappy, then, when, a few days later, Father McCallion came to the door and stood there talking to the teacher rather secretively. They were pointing out this and that and looking at us. The next thing we knew several of us—maybe five—were told to pack up our belongings, leave our books in our desks, take our coats and hats, and file out into the hall. It was all very mysterious. But into the hall we went, only there to learn that, after two weeks in the fourth grade, we had all been promoted. The fifth grade teacher was waving us into her room.

Other than Papa's having to teach me long division—he soon saw when he supervised our homework at the dining room table every night after supper that I was skipping the arithmetic problems requiring long division, a mathematical function we hadn't covered in the two weeks I was in the fourth grade—I did well in fifth grade. "My papa majored in mathematics at Rutgers," I proudly told the teacher when I showed her my fully completed arithmetic homework. "He even wrote a mathematical thesis on the pendulum!"

In short order I came to love St. Pete's, as we kids affectionately called the school. Around the top of the blackboard in every room appeared printed quotations readable by that particular grade from the work of some great authors or the speeches of great men and women. Oh, how I loved those quotations. I'd memorize them throughout the year. Even now I often find myself chanting a favorite from the fifth grade—"I had six honest working men; They taught me all I knew; Their names were What, Which, and When, Why, How, and Who."

I began to take learning very seriously. Books offered new things. I delighted in written exercises and spelling tests. Arithmetic remained a chore—I never imbibed Papa's love of math—and geography was fascinating if puzzling. It was hard to imagine a world you had never traveled.

Between Papa's political affiliations at home and the patriotism instilled at school—our national anthem, "America the Beautiful," and the Pledge of Allegiance deeply moved me—being an American began to mean something to me. And the intriguing world of symbolism was opened up to me first by a hatchet and two cherries.

———

The parish priest and school leader was Father McCallion. What a visionary that man was. He instigated and promoted wonderful activities in the school. We actually had an intercom system that connected the classrooms—this was in 1928!—and this intercom was connected to the radio. Every week the entire school would get to listen to the radio program by Walter Damrosch that had to do with orchestra and good music. I remember that Mr. Damrosch would describe the instruments, one at a time, and then you'd hear them altogether—all this in preparation for whatever orchestration was being featured that day. We all looked forward to that program every week.

By the time I was in the sixth grade I could play well enough for Sister Martha Mary, my piano teacher, to give me a quite difficult march (one I had to figure out for myself—the tricky rhythm and timing) which I was to begin playing each day as the students came up the stairs to go to their classrooms. The students would come up the stairs, walk in rows down the middle of the wide hallway, turn around at the end of the hall where they separated partners and walked individually back down against the walls of the hall to go into their classrooms. All the while they walked, the students were supposed to keep time to my piano music.

One day I had by now gotten so tired of "The Triumphal March" that I decided to divert myself by having a little fun. I had been told to keep the march to a certain speed, and the students had been told to keep time to the music. But on this particular day I began first to play the march very slowly, and the students would just be dragging as they went by the door. Then I would speed up, playing faster and faster, and the children would just be flying by the door. In a few minutes I looked up from my antics to see Father McCallion standing in the door, a little smirk on his face. He didn't say a word, but I immediately found the right tempo for the march, a tempo that from time forward never varied.

When the first bell rang—the school bell was a big electric fire bell which had a long ring—everything stopped as part of a very special ritual. Here were the rules: The minute you heard that first bell, no matter what you were doing, you stopped dead still. There was always much activity going on on the playground—kids throwing balls, jumping rope, skipping hopscotch, building human pyramids, spinning each other around like tops—but no matter what you were doing you stopped in motion the minute that first bell sounded. A pitcher would halt in midpitch; the person just spun would hold the position in which she had just landed; the hopscotcher would maintain his one-legged stance. No one would move until that second bell sounded. It was a sight to behold and widely discussed in the community. In fact, on most days there would be a covey of cars parked along the fence of the playground near the end of recess time, their occupants waiting to see the spectacle of the St. Pete Playground Freeze.

Of all my teachers at St. Pete it was Sister Vincent Marie who had the greatest influence, both scholastically and religiously, on me. My teacher in the seventh and eighth grade, she inspired in me a deepened love for academics and was the source of my whole outlook on sisterhood. Sister Vincent Marie was everything that I would ever want to be; I aspired to emulate her. She was a very fair person and a very good teacher. She also had a great sense of humor.

One day during our history lesson, Sister Vincent Marie suddenly called out a boy's name—I suppose he was misbehaving—and said, "Would you please, I want to see you after school." Under his breath—he certainly didn't think he was speaking loudly enough for the teacher to hear—the culprit responded, "Give me liberty or give me death." Sister Vincent Marie, looking over her shoulder from where she was writing on the blackboard, immediately asked, "Who said that?" Someone from the back of the room chimed in, "Patrick Henry." Well, this retort really was funny and Sister Vincent Marie roared over it. When we saw her laughing, the rest of us joined in. All concern for the boy's misbehaving was lost in the rollicking merriment.

Sister Vincent Marie was a stickler for grammar! Many times we would be out on the playground and someone would ask: "Can I get a drink, Sister?" She would always reply, "You can." Then when the person would start off, she would say, "But you may not." It took us a while to figure out why she was saying it like that, "You can, but you may not," but we soon realized that it was a grammar lesson in disguise. I also remember she used to say, "Stop slurring your words; it is not a never sharp pencil; it is an Eversharp pencil."

Because of Sister Vincent Marie I came to love English. When I graduated from the eighth grade, I received a gold medal. On the front the medal read *Written English* and on the back was my name. I wore that medal on a chain around my neck all during the ninth grade; I was very proud of it.

I wanted to emulate Sister Vincent Marie's devotion and piety, so I began taking part of every lunch hour to go pray in the church. Father McCallion had built a little prayer corner with statues in it off to the side of the altar. Wrought iron grillwork separated the prayer corner from the rest of the church and two small wrought iron gates opened back for easy access.

When I entered the little prayer corner during lunch every day, my eyes immediately riveted on the beautiful life-sized statue of St. Theresa of Lisieux, known as The Little Flower. St. Theresa was a Carmelite nun, and she was depicted carrying a spray of roses. The eyes of St. Theresa were glass—the statue was a lovely antique from Europe—and so beautiful. They seemed to look right through you. I would kneel in front of the statue, thinking about the life of St. Theresa which I had just read in Sister Vincent Marie's class and asking the saint to help me with my vocation, for I knew that I wanted to be a sister. It was encouraging to read that St. Theresa had entered the convent when she was only fifteen years of age; I knew she understood, then, the desire of a young girl like me. Only after I actually entered the cloister in New Orleans did I realize that the Carmelite Order to which St. Theresa belonged was also a cloistered Order.

Sometimes when I knelt there I would think of the statue of the Virgin Mary that Mama had on the little altar in her bedroom alcove at home, the altar where Sis, Bruds, and I knelt every night to say our prayers with Mama. The lines of some of the verses in *The Miraculous Medal Magazine*, a little publication promoting devotion to Mary Immaculate that we children so eagerly awaited every month, would come to me during that lunchtime prayer:

The King's Highway

I saw her walking through the field,
   God's Mother with her Son,
And every little flower-bell pealed
   To praise the Holy One.

And every lily lifted up
   To see the wondrous thing
As bearers of a dew-filled cup
   Before the little King.

Oh, every little rose upturned
   To wave as He did pass,
And every little sunbeam burned
   Its incense on the grass!

Oh, every little piping bird
   Did trumpet from the tree,
And every little lambkin heard,
   And danced, God's Lamb to see!

> Oh, nature all did serenade
> God's Mother and her Son;
> And then I knew why God had made
> His creatures—every one!

In this lovely quiet chapel everything seemed to fall into place. My kneeling in front of St. Theresa and asking her help was no mere sentimental episode, though to deny emotional involvement would certainly not be true. I wanted to do something for God and wanted to belong to Him in a special way. My relationship with God was a very private affair which I hesitated to discuss. (Because of my active love of life, nature, and the outdoors, the path that I would one day follow came as something of a shock to friends and even to my family.)

Before I left the prayer corner each lunch hour, I would move over from the statue of St. Theresa to kneel in front of the enormous crucifix that was in the same area. Sometimes when my eyes were affixed to the statue I would be transported by memory to Christmas mornings at home . . . .

Sis, Bruds, and I getting up from our beds in great anticipation of what lay under the tree for us. But at the same time knowing we had to obey the rules. "Walk, don't run," Mama would say. And down the stairs we would walk. Before we ran to see the tree and receive our presents, the three of us always moved slowly to the center of the room, kneeled down, and looked up at the wall to view the beautiful crucifix which hung between the two bay windows. The windows were a kind of stained glass with pretty crisscrossed metal on them, and there in the center between these windows hung the crucifix. "Our Father who art in heaven . . . " Sis, Bruds, and I repeated. And then we'd conclude, "And thank you, God, for coming into the world on this day and for dying for us." As much as the lovely tree and the wonderful gifts, the crucifix was synonymous for me with Christmas mornings.

Kneeling now as an eighth grader, remembering those earlier moments, I would look at the crucifix and repeat the Prayer to Christ Crucified which appeared on a card behind glass on the top part of the kneeler.

Look down upon me good and gentle Jesus while before
Your face I humbly kneel. With most fervent desire of my
soul, I pray and beg You to impress upon my heart lively
sentiments of faith, hope and charity, with true contrition
for my sins, and a firm desire of amendment, while with
deep affection and grief of soul I ponder within myself and
mentally contemplate Your five most precious Wounds;
having before my eyes that which David spoke in prophecy:
"They have pierced My hands and My feet; they have
numbered all My bones."

I could not remember, kneeling there praying that prayer, a
time in my life when God was not included. I had grown up
believing the truths of our faith. "Why did God make you?" At
age six, I memorized: "God made me to know him, to love Him
and to serve Him in this world and to be happy with Him forever
in heaven."

Early the questions and answers in the Baltimore catechism
had become my guide in understanding the teachings of the
Church and my help in everyday living. This catechism taught
and explained to me the Ten Commandments of God, the Six
Commandments of the Church, the Beatitudes, and the tenets of
basic theology. And while even now I did not understand, although aided by Mama at home and the sisters at school, everything I had committed to memory, I did have an appreciation of
God's beauty in everything and knew that sanctifying grace is an
abiding supernatural gift by which we were cleansed from sin
and made holy. I had known since at least the age of six that to
be in the state of grace meant to be free from mortal sin.

And I *knew* that prayers were answered.

Why, hadn't Mama and I gone over to see Mrs. Ake, our
dentist's wife, one morning . . . .

*Mama had put me in the back of the car saying, "Now, I
have something to do. I'm going to go see Mrs. Ake and see if
I can get her back in the Church. She hasn't been there for
years, not since she and Dr. Ake got married." When we
drove up to the house and parked, Mama had said, "Nina,
you sit out here in the car and you keep saying Hail Marys
and you keep praying for Mrs. Ake. I hope we will be able to
get Mrs. Ake to come back to church." I took Mama's
instructions seriously. To me it was a great commitment,
something that Mama had asked me to do and since it was*

something I could do and wanted to do, I sat there . . . and how seriously I prayed. Mama was in Mrs. Ake's house for what seemed like weeks, but when she finally came out, she was grinning and Mrs. Ake was waving. When she got in the car, Mama said, "Mrs. Ake is going to come to the service with me tomorrow night."

Unbeknowst to Mrs. Ake, Mama went to the priest and informed him who Mrs. Ake was and how she had been away from the Church for a very long time and how fearful she was about coming back. The next night when we all went into church—just as we entered the sacristy—the priest put out his arms, embraced Mrs. Ake, and said, "Welcome home." Mrs. Ake left that night, informed Dr. Ake she wanted to be remarried in the church—he surprisingly said, "Sure"—and Mrs. Ake was able to receive the Sacraments and return to the Church.

So, with experiences like that in my background, whether I was praying now to St. Theresa for help in my vocation or to the Crucified Lord for forgiveness and mercy, I knew that my prayers would be answered.

Leaving the chapel one lunch period I ran into Sister Vincent Marie in the schoolyard. We were standing around waiting for the bell to ring—I don't remember the whole of the conversation—but I do remember saying, "Sister, I think I have a vocation to be a sister." I have never forgotten her reply: "Nina," she said, turning to look at me, "just remember a true vocation is always silent." I decided that what she meant was that you didn't run around telling everybody, "I'm going to be a sister. I'm going to be this. I'm going to be that." So I decided for myself, "If a true vocation is always silent, then I won't talk about it anymore to anyone."

It was interesting that the next time I ever spoke to Sister Vincent Marie about my vocation was after I entered the convent in Allegany. She came to the Motherhouse one summer, and I was allowed to go out in the yard and walk with her. We went over and sat down on a bench. "Nina," she said as soon as we were seated, "When I looked up and saw you coming down the stairs dressed as a postulant, I was so surprised. Why didn't you tell me that you were going to enter the convent?"

I replied, "Well, Sister Vincent Marie, don't you remember that time on the playground when I told you I thought I had a vocation and what did you say? You said, 'Nina, just remember a true vocation is always silent.' So I had a true vocation, and I didn't talk to you or anyone else about it anymore."

Sister had seemed chastened by what she had heard.

Here I am, Nina Henriette Muller, age three. These are the high-top shoes which needed a button hook to fasten.

My father, my mother, and my sister Mary.

This is Mama in her seal-skin coat with the black fox collar and ostrich plumed hat. Corsages were worn at center on the lower part of the collar.

Sis at age fourteen with poodle in her lap, me at age seven, and Bruds at age three—all sitting on top of a pile of corn-stalks from Dr. Ake's farm. Sis is driving her pinto pony, Harry.

Easter Choir at St. Peter's Church, Pleasantville, New Jersey 1936

Top L to R: Jerry Vigue, my father George H. Muller, Pete Quenesso, William Schwartz, Tony Perri, Edward Stecher, George Flannery, Sr.
Third Row: Marian Nealy, Ms. Kitchen, my mother Nina Landis Muller, Annette Franklin, Ella Vigue, Bess O'Brien, Jule Martens
Second Row: Helen Smith, Agatha Marshall, Helen Owen, Mary Rogers, Antoinette De George, Agnes O'Brien
First Row: Rev. Diehl, Rev. Francis J. McCallion, Rev. Alfred Jess, William Flannery

All decked out in my sister's organdy dress, I am ready to go to the dance with Pete Starn. Photo taken by my mother.

This photo was taken as I had my last ride on my horse Zeb, July 20, 1934, right before I entered the convent. I was just sixteen years of age here.

This photo of Mama and me was taken on July 20, 1934, two days before I entered the Franciscan Sisters of Allegany.

Mama took these two photos of me at the grotto and the Way of the Cross statue in Allegany when I was a postulant there in 1934.

This photo of our family was taken at the Motherhouse of the Allegany Franciscan Sisters on August 16, 1936, the occasion of my profession in this Order of teaching sisters. Left to right: my mother, Nina; my brother, George (Bruds); me, now known as Sister Marie George; my sister, Mary Muller (unmarried here); and my father, George.

I was now a professed Franciscan sister.

SECTION THREE

NEW ORLEANS, LOUSIANA
1951-1959

In the Cloister With Much Reminiscing

CHAPTER FOUR

BINDING MY PAST LIFE WITH THE PRESENT

When I see your heavens, the work of your hands,
the moon and the stars which you arranged,
what is man that you should keep him in mind,
mortal man that you care for him?
FROM PSALM 8

May 10, 1951
New Orleans

Dear Sis,

Thank you for the check. My, but it is a big one! The thorn bush arrived safely. It is already in the ground. Ben planted it while Rev. Mother and I looked on. The moss around the roots was quite wet yet. The leaves had withered but I doubt that will make any difference. It is in a nice place with plenty room for expansion. It has just a year and half to grow, you know! The big sprigs are stiff. I am going to soak them in water tonight to see if they'll bend without breaking tomorrow. What is the correct name for that tree? Ask Helen and tell her thanks ever so much.

I haven't mentioned the weather as you say. Sis, it is ideal right now — breeze and all . . . but! as for August — I asked those who know — they said it is inferno. If I were you, I'd bank on October as you did before — even September they say is terrific and it would probably not be pleasant for you at all.

As for your writing to me — Rev. Mother says she enjoys your letters too. But of course you don't expect me to be able to answer frequently.

We expect to have the View Master for Pentecost Sunday. We had it all fixed up — chairs and all last week and word came for us to go to the infirmary. Five minutes after we arrived Mother Seraphina (a former Rev. Mother) passed away. It has been expected. She'd received Extreme Unction several days before. It was a beautiful death. We were all there saying the prayers for the dying. She was laid out in the chapel with the grate open — a Solemn high funeral mass. The priests and pallbearers came into the enclosure as we walked in procession to the vault out back. In case you are wondering, she was embalmed and buried in a casket — a crown of roses on her head. I have certainly had experiences here fast!

Enclosed find a booklet Mother Mistress gave me to send you. The photo in back of the garden shows only a small portion of it. That reminds me — Rev. Mother said when you come down you could bring the camera and attachments and we (not you!) could take some photos for you on this side of the wall! Speaking of coming down, Mother Mistress just told me to tell you it is hot here in August!

Enclosed find a Sacred Heart badge — the first I crocheted — not too good but I wanted you to have the first, a Mother's Day gift — ha! Will remember you specially on Sunday. It is a consolation to know too that Father Juvenal is taking care of Muds' monthly mass. When did we start them? September?

Tell Dr. Vettese we have a dental chair in the infirmary. They say the dentist is very good and obliging. The nuns' teeth all look lovely to me. I gave Rev. Mother that impression Dr. Vettese gave me to bring along. Here's hoping we never need to use it.

I certainly have gotten use out of the pen and pencil you and Papa gave me. The pen hasn't been refilled yet!

Rev. Mother let me read the letter you wrote to her. You make me smile, Sis. You say you've never done anything for me. You goon! What more could any sister do for another? I could never repay you for your unselfishness and devotedness and you know that.

Rev. Mother gave me permission to write to Sr. William and I will just as soon as I have a chance. You will certainly have a lot to tell her when you visit her next. I know she'll love to get the glad tidings from you.

Today, on the altar, from our own garden, Sister arranged beautiful bouquets of white and pink glads, white daisies, and calla lilies. The oleander trees are super . . . red-red and even fuller than that Japanese Snowball we used to have when we lived at 226 West Jackson Avenue.

We're anxiously awaiting the screen. I know it will be enjoyed. I'll write out a list of the numbers of the reels we have for you sometime. I really haven't had a chance to do that yet.

The books you bought for me are certainly a practical gift, Papa, and the music, too. None of it was duplicated here.

Well, Happy Mother's Day now and thanks again for the bush and the check. That is a treat.

Oh, I forgot to tell you. I'm going to have a room to spread out in — my art work, I mean (ha!) where I can leave it undisturbed. Love to you & Pop & Bruds and Nina Marie and Kathy.

Nina

Thanks for the stamps.

It was in the garden and yard of the monastery that I spent much of my time in the early days of my life in New Orleans. The property took in a complete city block, and the grounds were extremely well kept. Although we had a wonderful black man, Ben, who was responsible for the grounds, we novices did a lot of garden work, also. One of my outside jobs was to pick figs. It was a trial to climb the ladder in my long habit to reach the figs in the tall trees, but it helped when I was struggling so hard to remember how good the fresh figs tasted every evening when we had them with thick cream for supper. Many times I had to battle the ants who seemed determined to get the figs before I did. I also helped the elderly nun from Germany harvest her "pookins," as she called them—various kinds of pumpkins and pumpkinlike squash which we ate in casseroles as well as pies. Strange to my eyes—and extremely pleasant to my taste buds—was a vegetable the sisters called a melatone, a sort of fluted, pear-shaped vegetable which hung down from very tall vines. These vines had to be surrounded by a cage-type contraption made of chicken wire to discourage the squirrels who would, if allowed, eat the young vegetables before they had a chance to grow. Oh, were those melatones good when they were boiled, cut in half, the flat seeds removed, then butter and cracker crumbs put on the open half and run under the broiler in the oven.

The grounds also contained a mausoleum that we called Calvary. Since there could be no underground cemeteries in New Orleans because of the high water table, the sisters were buried in this small building that had shelflike places for the coffins, with marble slabs in front of them. Early on, I picked out my gravesite right near the front door so that on the last day I'd be ready to get up and go!

I felt fortunate to have come to a monastery whose enclosure was so beautiful. There were pecan trees, oak trees—ancient live oaks with very tall, large trunks and many of them, beautiful with the long Spanish moss hanging from them. The azaleas in the yard were breathtaking, and there were flower beds everywhere . . . shasta daisies and beautiful poinsettias which Sister Mary Agnes had as her own special domain.

Every time I was in the yard my eyes sought out the great big hydrangeas. As I had told Mary when I wrote her, these beautiful balls of white did look like the Japanese snowball bushes we had back on West Jackson. Sometimes during quiet time I would come out in the garden and sit on a bench remembering my childhood home place. . . .

There were three front gates to our wonderland: One opened to a sharply-edged, manicured path winding through shrubbery to either the back door of the main house or branching off on the trail past the pond to our summer cottage way in back. The center gate, through which most visitors entered, led abruptly past the sky-blue spruce tree directly to our front porch door. The double gates opened for the car or pony cart and served as a seat, if a precarious one, from which one could view a very placid outside world. (Mama liked for us to stay always in our own yard—"Where else is there to go that is more beautiful?" she would ask, and we had to agree.) In fact, I would wonder as I sat in the monastery garden thinking of those times why I had ever bothered to sit on that fence facing the road when so much beauty was behind my back!

Towering oak trees protected most of the acres from the summer sun. Our yard was a cool place to be. It was nice to sit in the shade where Papa had built several permanent seats encircling the fused trunks of what appeared to be not four but one huge oak tree. Better still was the secret cave of hedge, good mostly for hiding out and just thinking. We loved a swing whose rope reached skyhigh, putting to shame the confinement of a pendulum in the grandfather's clock.

Lush green lawns were mostly for beauty but felt so good on those rare occasions when we were allowed to run barefoot around the sprinkler on a hot day. A roof-high wall of hedge, bordering the entire back yard, offered privacy from the neighbors. But that hedge had other purposes as well. That was where Mama sent us to get a switch when we needed whipping.

Papa must have thought about planting fruit trees long before contemplating me! For as long as I can remember there was an abundance of summer and fall fruit: pears, peaches, cherries, quince, and plum, plus huge bunches of big purple grapes.

But it had been Mama's scheming that had brought beauty in color. Hills of rocks spilled over with colorful crawling flowers while her outdoor room of rustic furniture was sheltered beneath a solid canopy of living wisteria vine. A shoe-shaped pond, teeming with golden flashes of fancy fish, harbored living bouquets of water lilies. It boasted a rustic bridge spanning the center, artistic to adult eyes, I am sure, but to us children a good seat for dangling hot feet in cool water or for teasing exasperated fish. All this beauty could be seen through the sun parlor windows of the house.

There were dahlias that grew as large as sunflowers. And the pure white perfumed carpet of lilies of the valley that grew under that umbrella-shaped snowball tree . . . while nearby grew Mama's pride and joy, the scarlet red Japanese maple tree.

"Yes," I thought, bringing myself back to the scene in New Orleans, "to be surrounded by so much of God's beauty in childhood was just a foretaste of what awaits us in the hereafter." These monastery grounds where I now sat—themselves a touch of magic—seemed a continuation of those childhood blessings.

May 16, 1951
New Orleans

Dear Sis,

Just received your nice long letter telling all about the housecleaning, arranging and flower planting. The place must look lovely. This morning I discovered four fig trees with small green figs already growing. The trees are large.

Well, to get to the real reason of this letter — the screen arrived! It is just like you to get the best possible! We were wishing it would arrive for Pentecost Sunday since we had a long recreation period in the evening. We looked at pictures for over an hour. I flashed them on the wall. Well, Monday the package from Sears came. We were able to sit in the library since the screen is portable. I thought it would be one that you hang up! What a surprise! We flashed some of the same pictures to see the contrast. What a difference! Sis, those colored photos are just gorgeous on the screen. Many of the nuns had never seen a black & white movie, let alone a colored film. We've seen about half of the reels so far. Enclosed find the numbers of the reels we have. I finally got a chance to write them out for you. I'm glad I brought the 3 Little Pigs, etc. The nuns enjoyed the dramatizing. I wonder if there are travelogues with all reels? Some had them and it makes it nice. I'm so glad we have the screen. Half the beauty was lost before! The nuns said for me to be sure to thank you.

While I think of it, will you write out your recipe for aspic salad, giving the quantity it makes. I can't remember it and I think the sister in the kitchen would like it.

The thorn bush looks dead but the stem near the trunk is still green. Should I prune off the apparent dead branches? Tell Helen to save all the branches she cuts from the big tree. I'd appreciate it if they could be tied in a circle. They could be mailed in a cardboard carton. Of course, I know she prunes in the fall, so I mean then.

This week I'm ringing the Angelus. Remember the time I rang it when you all were in Allegany?

The weather is lovely today. I can't imagine it being terribly hot but they say it will be. I'd advise you to come later rather than August. If Papa can't make it till later I think that would be better. I'll have the holy habit by then, please God.

Today I am one month old. It seems the most natural thing to be here. I've never felt so "in line" if you know what I mean. I am actually relaxed and can be myself. Living from day to day brings all grace necessary.

Have been enjoying the letters from Papa and, of course, from you and Bruds.

Must go now to ring the bell for meditation and rosary.

Thanks again and again to "you all" from "us all" for the wonderful screen.

Love and prayers,
Nina

If you come across any of those flat rayon or nylon shoe laces, any color, we can always use them for bookmarkers. Those wild new colors are only 10 cents in the 5 & 10 store.

Send the aspic salad recipe . . . tell the dentist this and that. . . . I don't think there is denying that, happy as I was to be with the Poor Clares in New Orleans, I also was homesick for my home and family. Was it because I knew I would never see those beautiful grounds again? Eat at my family's table? Talk to persons in the outside world unless they came to stand at the grill which demarcated the enclosure? I often found old memories flooding over me.

There had been that first night after arriving when I could not sleep. First I had thought it was the hard bed; in fact, I remember waking up once and thinking, "You've got to get up from off this floor and get back into the bed," only to realize when I became fully awake that I was *on* the bed—but it was just as hard as the floor. It wasn't the lack of comfort of the bed, however, that was my problem. There was the ding-dong that kept ringing right outside my door, a sound that for the life of me I could not stop thinking was the sound of the elevator at Blatt's Department Store where Mama used to take me on special trips when I was a little girl. Ah, those memories. . . .

"Nina, what kind of parfait are we going to get today?" I could hear Mama's voice as we got off the trolley and headed for the department store. "Butterscotch or chocolate or strawberry...or perhaps they might even have something new today." Such delicious agony, to decide what flavor parfait we would have when we finished our shopping and stopped in at the little restaurant on Blatt's mezzanine.

The year Mama and Sis went Christmas shopping together. So we parked out in front of Blatt's and Sis said to Mama, "I'm going to go in first and get your present, and you stay here." Mama and I sat in the car and waited. Sis went in and came out with this big box. And then Mama went in. Lo and behold, she came out with a big box just like Mary's. They laughed so hard because they could tell by the size and shape of the boxes that they had bought each other the same thing—a comforter for the bed. So they just opened the boxes right there in the car and each one decided she liked the quilt she had bought better than the one the other had bought for her. Each just kept the gift she had bought! They didn't even bother to wrap them and put them under the tree. We laughed for years about that Blatt's shopping trip.

One day soon after my arrival at the monastery, I was working in the kitchen. Crash! The wet cup slipped through my hands and fell to the floor before I even knew what had happened. "Sister Henrietta, pick those pieces up right now and take them to the mistress of novices!" said the nun in charge of dish-washing. "You will have to do penance!" "Why, I didn't have to do penance even when I dropped that whole tray of glasses the first week I was in the convent in Allegany and broke every one of them!" I thought to myself. "And now here in New Orleans I'm going to have to do penance for breaking one cup!"

"But," a rational second voice in my head replied, "you were sixteen when you entered the convent in Allegany and people excused you, but it's a different story now when you are thirty-three. And, besides, you know that doing penance for small things is an effective means for guarding against future mistakes by accepting with humility whatever hardship the penance entails."

The piece of the broken cup containing the handle was tied by a string around my neck. "You will go, as all others do when they break something, and kneel in the refectory," the mistress said, "wearing this sign of your carelessness. When the other sisters come in for lunch, they will see you and be reminded."

I obeyed the mistress and went to kneel in the dining hall, the cup shard dangling on a string around my neck, my eyes closed to everything around me . . . .

Mama and I are in Blatt's again. I must be about six years old. I am following Mama as we walk through the china department, and somehow just as I pass it, a set of china on display in the aisle crashes to the floor. Plates, cups, saucers, and bowls fly in all directions. I am sure the sound of dishes breaking could be heard on every floor. When the department manager tells Mama she will have to pay for the entire set of china, she was so disturbed that she won't even take the pieces that are not broken. After paying the bill, Mama, without one word of chiding, takes my hand in hers and we walk toward the elevator.

This year I was eight. How often we heard Mama tell the story: "Two Japanese men came to visit my family when I was a little girl," she would recall. "One was named Hiranuma and one was named Mitsui. They gave my mother this bowl and also those long strips of incense that we still use to keep the mosquitos away when we are outside at night sitting by the fishpond." (Punk, Mary, Bruds, and I irreverently called it.)

Well, one day I don't know what happened, but I broke that precious bowl. It fell to a million pieces on the floor. I was frightened and didn't know what to do, but since nobody was around I hurried and picked up the pieces, one by one, took them outside, and heaved them over the fence at the back of the property where they fell out of sight among the leaves and huckleberry bushes. The next morning at breakfast, Mama said, "Whatever happened to the Oriental bowl? Where could it be?" I thought, "Oh, oh, do I tell her or don't I." I was in such a quandary. I knew obedience was the highest service to God . . . hadn't Mama always told us . . . so, finally, I answered. "Mama, I broke it." Mama replied, "Yes, I'm glad that you told the truth. I know you broke it." "How did you know I broke it?" I answered, because I knew no one had seen me. "Look under your feet on the floor." I looked down and there it was—something I had overlooked: a great big piece of the Oriental bowl.

Now as I knelt on the hard wooden floor in the refectory in New Orleans, I wondered: "Why can't they respond to this broken cup the way my mother responded to that shattered Oriental bowl?" Her method had made a much greater impression.

Mother Superior, taking note that all the nuns had now marched in and were seated ready to eat, began the noonday prayer. When she finished I would be able to rise and take my place among the sisters at one of the long tables. But as she prayed, I remembered another broken cup experience I had had with Mama—this time when I was a teaching sister in Ft. Myers, Florida. Mama and Sis have come from their rented winter cottage across the street to eat with us sisters at the convent....

Mother Lucian is gone for the day—did she drive over to Tampa on business or maybe to Miami to visit the nuns at the Hospital of St. Francis?—and Sister Conrad, Sister Loretta, and I were feeling "full of it" with our Superior out of town. The cups and saucers we had were really cheapy— kind of ivory color, nondescript kinds of cups that you buy at the five and dime—and during the course of dinner Mama accidentally dropped her cup and broke it. Sister Loretta, I suppose to relieve Mama of the burden of having broken the five and dime cup, picked her cup off the table and dramatically held it off to the side and let it go.

Well, when that cup hit the terrazzo floor, there was no chance of survival. I don't know what got into us, but we all took one look at Sister Conrad and she picked up her cup, dramatically held it out to her side and released her fingers. Another catastrophe. It was now my cue, and I did likewise. We ended up without a single cup on the table.

It was only after the spontaneous fun had ended that we looked at each other and said, "What are we going to do? Mother will be coming back and there won't be any cups." I said, "Don't worry, we'll go down to the five and dime—I saw the same kind of cups there for a nickel—so we'll just go down there and get them." Mama said, "No, I'll do it."

So she went down and got the cups—only the shape of them was a little different. It was so funny to see Mother Lucian's face when, her first morning back, she sat down at the table. She looked around from one cup to the other and just could not figure it out. I don't think anybody ever told her why we had new cups. We had got such a bang out of it.

I am sure that none of the Poor Clare sisters in New Orleans had an inkling of why I was smiling when, upon the Mother Superior's beckoning, I rose from doing penance on the floor of the refectory and started toward my place at the table.

Early on, Mother Superior gave me my own little garden spot in the enclosure. I liked getting out of doors and took immediately to the task of raising flowers. Zinnias I raised from seeds; how much I loved first planting the seeds and then watching the little plants grow until there were beautiful flowers I could cut and put on the altar.

The yardman, Ben, helped me plant a twenty-five-foot row of long-stemmed early spencer sweet peas which we also trained to grow up over a lovely arbor. What a profusion of splendor these flowers made! The stems were very long and the flowers grew in every color imaginable. I cut the sweet peas, too, and put them on the altar.

My next project was planting morning glory seeds. These I soaked overnight in warm water to soften them a bit and help speed up the germination. Soon I had a garden plot of heavenly blue morning glories. Observing how they budded and bloomed, I got the idea for playing a joke on the sisters when we went to pray before the altar. I devised a tiny vase just large enough to hold one morning glory stem. Then I placed dozens of these little vases around the altar, each holding a bud I had just cut that morning. I knew the buds would burst into magnificent bloom sometime while we were praying, and that was the joke I wanted to play on the sisters. When they closed their eyes, nothing but buds on the altar; when they opened them, beautiful blue morning glories.

The Mother Superior's response particularly tickled me. "I was so surprised and shocked," she told me after the service. "When I closed my eyes there were only green buds on the altar, and when I opened them, beautiful blue morning glories. Until I figured out what had happened, I thought there must have been a miracle!"

The joke was on me, however, when a few months later, the night-blooming cereus placed on the steps of the altar bloomed just at midnight, at the moment we were saying our office. The bloom was gorgeous and lasted only a short period of time. Now that did seem to me to be some kind of miracle, that a plant which was due to bloom only one time in its life and then only for a few hours would burst into blossom at the moment we were praying.

The only real disappointment I felt in my choice of life here with these sisters was the weather. When the cool days of May passed, I came to know the true climate of New Orleans: every day was hot and mucky, depressing kind of weather. When we prayed, we knelt in pools of our own water. I remember being slippery all the time. We wore cotton shirts with little sleeves under our serge dresses, and these shirts just soaked up the perspiration. The serge dresses were hot in themselves but on top of the dress when we went to chapel we put this cape made of the same hot material and, of course, always there were those little hot serge bonnets that certainly didn't add to our comfort. To accept this as penance was the only thing to do. I never felt particularly annoyed by the situation; it was just all part of being a Poor Clare cloistered nun. But this didn't keep me from day-dreaming sometime in chapel. Instead of the sound of the mass I'd hear the lazy slapping of the waves as they hit the beach when I was still a teaching sister in Ft. Myers, Florida. . . .

Mother Lucian was very wise. We worked hard and willingly at teaching all week, and by Friday we were eager to pile into the car with our bathing suits and beach towels and head down the trail toward the pink cottage at the secluded end of the beach. (This was the same cottage that we prayed so assiduously for once during a hurricane, discovering when we were finally able to go check on it that the cottage had been picked up, turned completely around and sat back down precisely on its foundation. The only difference was that now the back of the house faced the ocean and the front door faced the backyard—all the pipes underneath had been snapped off of course, the double-car garage was completely gone—we couldn't find a stick of it, the furniture was scattered around a little but the strangest thing was that on a shelf in the kitchen there were several glasses and an open bottle of Coke. Not one drop of that Coke had been spilled and the glasses were all still in one straight line. Not a single one of them had been broken!)

It felt good to rest the week's weary bones on the warm salty sand down by the water in front of the cottage. I always envied the starfish on the beach. They were able to look the sun in the eye. But when the sun got too intimate we'd make a dash for the waves. Their constant welcome was hard to resist. We diluted all the care of the week and washed away every problem. Our laughter would be drowned by the cry of the gulls as they announced the quiet approach of the porpoise.

We'd strain our eyes to count their bobbing heads as they washboarded by. We were never afraid to share the water with them. They were appreciative companions to us humans. Somehow we always felt safe because of their presence. I envied the porpoise. We finally had to leave the sea. They never knew the burden. The buoyancy of the salt water spoiled me as the sand once more beneath my body called me human. But it was a nice time to be human! My feet always enjoyed the smushy sensation of the wet sand on the beach and I'd watch my toes exult as the wet sand came oozing through.

The warm sea breeze had a taste of its own, like an hors d'oeuvre preparing us for a good pick-up up at the cottage. We always brought food with us that needed very little preparation.

No television or radio interrupted our tranquillity. We lived the good life and loved every minute of it. I remember only Chinese checkers to tax my mind. We read all the books we never before had a chance to open. The beach experience was a retreat for mind, body, and soul. Many, many an evening I watched the sunset over the Gulf of Mexico cast its fiery reflection heavenward and the glowing kaleidoscope of color sizzled its name upon the sea. Those were the times when the only dreams were daydreams, and night held things not to be remembered. This is the Florida I knew and this is the Florida I shall never, never forget.

Then a movement of the priest or a sound from the organ would bring me back from my reverie, and I would realize that I was kneeling in a pool of water in New Orleans, not hearing the waves on the beach in Florida.

CHAPTER FIVE

PRAYER, YES; BUT WHAT ELSE GOES ON BEHIND CLOISTER WALLS?

There are different gifts but the same Spirit;
there are different ministries but the same Lord;
there are different works but the same God
who accomplishes all of them in everyone.
FROM I CORINTHIANS 12

My days were full at the monastery; projects were many. From my arrival in April until I received the holy habit in November, I planned and worked on my wedding dress, which I made entirely by hand. I found everything manageable except the sleeves, and for these I had to ask for the help of Mother Margaret Mary.

The twenty stations of the cross needed refurbishing. What intricate work it was to remove the paint from the twenty plaque-type statues with their ornate and complicated figures and then airbrush each with fresh color.

This very large room on the second floor, which had been previously a sewing room, Mother Margaret Mary gave me for my art work. The room contained a long row of windows on the inner side which, if opened, would have provided a sight of the major chapel; but, of course, I was never allowed to open those windows. There were built-in cabinets to hold supplies, a long table, and a kind of drafting table with a glass top and fluorescent lighting which could be used for tracing designs and seeing through paper.

One of the first uses I put my new art room to was painting stoles after Sister Mary Roch had made them. These stoles, which were used during mass, were made of white moire or more often white satin. My task was to airbrush colorful designs on the stoles. I also painted burses and other items used on the altar, such as ciborium covers. And tabernacle veils . . . ah, how many of those I painted.

Then Mother Margaret Mary suggested I take up oil painting. Just as with the stoles and veils, I had to figure everything out on my own, reading books on the techniques or trying to learn through experimenting. I first had to learn how to stretch my own canvas and then learn how to paint on it. "Why don't you do portraits?" Mother asked me after she saw that I was getting the knack for this kind of painting. So word went out and people began sending in their photographs. At one point I counted up and I had already completed sixty-two oil portraits, including that of the doctor who had to come to the monastery when I couldn't get his likeness just right from his photograph. Mother Margaret Mary had him come sit in the parlor while I, on the other side of the screen, sketched him. And one of the benefactors even asked me to paint his boxer dog! Another wanted a large painting of magnolias and sent in a big box of live magnolia blossoms as models.

In 1954 the Pope named Holy Mother Clare the patroness of television in honor of a vision she had once been granted. It was 12:00 midnight, Christmas Eve several hundred years ago, and Mother Clare, sick and unable to leave the monastery to attend mass at the church in the nearby village, had nevertheless been able accurately to report to the sisters when they returned every part of the service. She had, through God's graciousness, been able to see the entire service from her sickbed many miles away.

When we heard that St. Clare had been given this signal honor of being named Patroness of Television, I told Mother Margaret Mary about my own favored experience—the time I was asked to play the organ for the first mass that had ever been televised in the United States. I had been in Haddonfield, New Jersey, at the time. I told her about the huge outdoor cables that led from the big television truck—one of only two in existence—the hot, bright lights, and huge cameras.

"Why don't you paint St. Clare's vision?" Mother Margaret urged me, excitedly. I obliged Reverend Mother, and when she saw it, she was very happy. "We will make a gift of this painting to Loyola television station," she determined. I was surprised at how everyone seemed to love the painting, because I was never happy with it—"too stiff" I'd say to myself every time I looked at it. No one else seemed to agree with me, however, and Mother Margaret Mary even sent a pen sketch of the painting, which she had me do, to Germany to have a medal struck. These medals she distributed to many people, including the Pope who had designated Saint Clare the Patroness of Television.

Try as I might, I could not seem to avoid upsetting the routine in the monastery. The disruption would happen in the most unexpected of ways. Mother Margaret Mary was always encouraging any talents we young sisters might have, so she suggested that a number of us get together and put on a play. Well, I decided during rehearsals that we had to have some crepe paper for decoration. So I made the request of our Superior, having no idea that not only was there no crepe paper anywhere in the monastery but that no one had ever even thought about sending outside the monastery for anything so frivolous.

Finally, the governing council was called to a meeting. Would the young sister be granted the privilege of obtaining some crepe paper? After much discussion and registering of concern, it was decided that Sister Mary Roch, one of the extern sisters, would go into town and purchase a fifteen-cent roll of crepe paper. When I see Sister Francis Clare who was one of my coconspirators then and is now head of the New Orleans monastery, we have a hearty laugh over the commotion that was caused by that request for a roll of fifteen-cent crepe paper.

But this request seemed to signal the beginning of a sort of revolution, a turning over of some of the old customs. We younger nuns had plenty of ideas, many things we wanted to do. And more and more Mother Margaret Mary went right along with them. Not long after the crepe paper incident, I suggested to the young sisters that we do something special for Mother

Margaret Mary's feast day, the Feast of the Sacred Heart. Mother Margaret Mary agreed to go along with it, although she had no idea what we would be doing.

"Let's have a rhythm band," I suggested to my confederates. Then I told them about the rhythm band we had in our family, how my parents would make all the instruments, including the big horn made out of a dish pan and tin cans. So once again Sister Mary Roch was sent out of the monastery in search of another unusual request . . . dish pan, big spoons, combs, thin tissue paper, tin solder, and noisemaking kazoos. When she returned, I took myself secretively to the monastery basement where I spent hour after hour working on the big horn that was to be the centerpiece of the band. I had no idea how Papa had made it, but I just kept on until I figured out how to do it, and I did it. Tin cans scavenged from the kitchen I first flattened and then soldered together to make the curving part of the horn. The dish pan I fastened somehow to the end to provide the tuba-looking part of the horn. Then I put a kazoo on the blowing end so that one could really play the thing. I tried to make it just like the one Papa had made and Bruds played.

In the long hours I would sit in the basement working on that horn I would think with such happiness of all the times our family marched up on the stage for the Saturday night revelry at the River View Rod and Gun Club, each of us playing our rhythm instrument as we marched . . . *the night someone dared Bruds to play the big tin horn, and how finally when the dare was accompanied with a twenty-five cent piece he stepped forward, wiped his lips, and began to play that big horn right on the spot...how we had a flat tire about 3:00 a.m. that next morning going home after playing our rhythm band. . .and how, when several other cars from the club stopped to help us, Mama and the rest of the women and girls sang while Mama danced in the middle of the road with Dr. Ake.*

This revolution of new ideas, however, took a turn that we young sisters didn't especially like when as a result of Rome's move to federate the Poor Clare monasteries around the United States, Mother Margaret Mary and Sister Michael left the enclosure to fly to New York to visit the New York monastery. That trip, within itself, was a big innovation. But they came back very enthusiastic, ready to make many improvements.

The first "improvement" we became aware of when we went down to breakfast a few weeks after the sisters had returned. There at our places, instead of the customary silverware we had been using, were wooden spoons and forks. Somehow among all the changes Mother Margaret Mary had witnessed in her travels, it had been the wooden spoons and forks that had captured her fancy. We all thought these utensils were horrible but we were almost afraid, after our own capers, to suggest anything negative about something Mother thought was an improvement. Later we did get a new organ, an idea that had also come to Mother as a result of her trip to New York, and we also began to hear more regularly from the other monasteries.

After the Poor Clare federation had been in existence for a while, I got the idea of beginning a communication system. Explaining this to Reverend Mother Margaret Mary, I was delighted to find her very receptive of my idea. "I'll ask all the monasteries to submit articles, news, poems, drawings, or whatever they wish. This way we will have more contact with other Poor Clares and see how they live and think."

Unity was the name given to the publication. I learned to type well enough to prepare the copy for each issue, and Sister Mary Francis Clare learned to operate the multilith printing machine which Mother Superior bought us. One of the newspaper companies in town donated an enormous old camera—it must have been at least fifteen feet long. We used this camera to make our own negatives which we needed to burn the plates for reproducing photographs for *Unity*. Every month until I left New Orleans, *Unity* went out to all the monasteries of Poor Clares, proving to be a vehicle which put the monasteries in touch with one another. Before that we knew very little concerning other monasteries.

The idea for doing the *Unity* publication stemmed from my preparation of a book to commemorate the one hundreth anniversary of the New Orleans monastery. Using my camera, the last gift Mama had given me before she died, I took many pictures of the nuns and the grounds and used these to make up a book. I was extremely pleased with the outcome, given that I had had to teach myself everything as I went. When the yearbook publisher came to look at the book and I had laid the entire thing out myself and every page was printer ready, he seemed amazed. "Why, I think we're going to be able to photograph this just exactly the way you have made it," he remarked. I must say that I felt a sense of accomplishment.

Things of a publishing nature seemed to have become the order of the day for me. Mother Margaret Mary had now placed me in charge of the library.

"I'm asking the contractor to add a big room in the new infirmary," Reverend Mother had announced during one of our meetings. "Sister Bernadette will be in charge of the new library." I took on the job with gusto. I didn't know a thing about the Dewey decimal system, but I arranged some kind of order for the books that I could explain to the sisters. Now, of course, that library has grown considerably and the books have been correctly catalogued.

At the time I took over and moved the books to the new library, I noticed that many of them were desperately in need of repair. This prompted from me a special request of Reverend Mother who, in turn, brought it up as an important matter at the next council meeting. "Sister Bernadette has asked us to purchase a big book-binding kit which she saw in the Gaylord Catalog," Mother reported to the council. "I didn't know if we could afford the $34, but Sister explained how much longer our books would last and how many we now have that are in poor condition but could be made like new with new binding." The justifications I had provided did the trick, and I soon had my kit.

When the kit arrived, I was amazed. Standing almost three feet tall, it opened up like a book and stood on a table. Inside little drawers and shelves were filled with all the materials one needed for book binding, with the exception of some boards which I had to have and some extra clamps used to hold the books tight together when you are working on them. There was an instruction book included, and, after a little experimenting, I was soon putting hard covers on soft-covered books, stitching up old books, and repairing the worn ones. I loved the work, and I must say it kept me very busy.

SECTION FOUR

FT. MYERS, FLORIDA
1940-1946

Life, All Sunshine and Music

CHAPTER SIX

MEMORIES OF FLORIDA NEVER
TO BE FORGOTTEN

I remember the days that are past:
I ponder all your works.
I muse on what your hand has wrought
and to you I stretch out my hands. . . .
It is good to give thanks to the Lord
to make music to your name, O Most High,
to proclaim your love in the morning
and your truth in the watches of the night.
FROM PSALMS 7 AND 19

When I first entered the convent at age sixteen, my duties and responsibilities focused on completing my high school studies and fulfilling the requirements asked of every postulant and novice. Upon taking permanent vows I left St. Elizabeth's and began a series of elementary school teaching assignments. Following these assignments I was transferred to Ft. Myers, Florida, where I was charged with teaching music at the local Catholic school. I had been on this assignment for six wonderfully fulfilling and happy years when a drastic change came in my life.

It was now the summer of 1945 and, as I had done every year since I had become a professed sister, I enrolled in the summer school program Reverend Mother Jean Marie suggested.

This year I was a student at Pius X School of Liturgical Music where I was working to advance my own performance and study of piano and organ. As I began my summer study program, never would I have guessed that my life would ever be about anything but teaching and music.

I enjoyed everything I was doing. And the study every summer always seemed to pay off, for the pupils I was teaching in Ft. Myers had become more and more accomplished and the recitals and concerts we gave, more and more professional.

Had I not been able, even within my first year at St. Francis Xavier, to get Ft. Myers named as a National Piano Teachers Guild Music Center, which meant our school was now a center for national piano-playing auditions? Even though we had only thirty-five students enrolled in our program, it had been the standards to which the students were held and the amount of musical materials they had studied that qualified us for this prestigious honor. Had not the master pianist himself, Hans Barth, come to hear my students and others play and to judge the piano-playing competition? Had not one of my students been chosen to be auditioned at the Juilliard Conservatory of Music in New York, and had not with such pride I accompanied him to that audition? And this spring concert . . . ah, had it not been the best ever, about fifty students, many playing complicated two- and four-piano pieces, with the entire Ft. Myers cultural and civic communities in attendance? And had all this not been climaxed by my receiving a personal note from Mrs. Thomas A. Edison praising my students? Yes, all was well as I settled into the comfortable life at Pius X School of Liturgical Music with my former novitiate companion, Sister Mary Cecelia, a very gifted musician.

Following the summer school term I went to New Jersey to spend the customary two-week vacation with my family. It was while I was at home that the bombshell came—a letter from Mother Jean Marie.

You have a new assignment. You will be moving from Ft. Myers to Bolivar, New York, to teach music at Bolivar Central School. You will go straight to the convent in Bolivar after your vacation in New Jersey.

This necessity to make frequent moves was about the most difficult thing I had found in religious life. At the end of each school year each sister was required to pack all her belongings into a traditional trunk. (Over the years, Mama had given me a whopper with two trays!) While at home on your two-week vacation in August, you might or might not receive a letter from the Motherhouse assigning you to Timbuktoo. You never returned to your former home; they just shipped your previously packed trunk to you. Then you went on to encounter a whole new set of faces, a new family, and sometimes drastic changes in culture, climate, and teaching assignments.

The sadness in leaving the group of sisters with whom I had lived for several years took the edge off any possible excitement of anticipating a new place. I just never got used to that; it tore me apart many times.

But I would never have thought of objecting. I knew God spoke to me through my superiors. Once I had to use firm measures to prevent the people in the parish from petitioning the Motherhouse objecting to my removal. I could never try to pry into the mind of God. He knows what He is doing with me and I am at His disposal. How difficult we could make it for our superiors who must make so many decisions! We cannot see the whole picture as they can. If this is so on the human level, how much more so is God in control. No, I had chosen this way of life and for better or for worse I would take whatever came with it. There is honest, legitimate satisfaction in really knowing your every move is actually the will of Almighty God!

But this move from Florida was the most difficult. After six years away from the home of my birth, I finally had found the beauty of family life in the convent. The group of sisters with whom I lived, our superior, and our pastor were truly united by a common bond which gave expression in abiding, understanding mutual friendship. Two of us were of German descent, one of Irish descent, and three directly from Ireland itself. We all just seemed to blend perfectly. Mama's "Blood is thicker than water" took on a new but true meaning. This family was different from any other; it had all the inner spirit of service to and for God, but something else was there! You could feel it from the start.

For one thing, we were not afraid to let our hair down. We enjoyed one another as friends and family. Kindred spirits had met and we knew it. There was a bond of loyalty, a sharing of the inner person as well as of material things. Somehow I sensed that this was the way religious life should really be; this is what I had longed for and did not realize it. This was a celebration of relationship you never wanted to see end.

Our relationship with the pastor was especially unique. There was a spirit of jovial banter alongside a great respect for our chosen vocations. It was an ideal situation which encouraged us to be ourselves. This pastor was an unusual man, a holy priest and so much fun to be with. He complemented the group of women with which I lived. It helped to make family spirit.

We seemed to be of one mind in the things we did or wanted to do. Of course, we did not always agree but it was easy to please one another. Our prayer life was good and we played well. Our community prayer was always pleasant and regular. With this family I learned to play, to relax, and finally to feel that I had at last "come home."

So not to go back to Florida?

To the beautiful white brick two-story St. Francis Xavier School with windows all trimmed in the most vivid red, surrounded by poinsettias which bloomed a good part of the year, big blossoms as big as a dinner plate, vivid red blossoms against dark green foliage; and facing fields and fields of gladiolus, flaming red, salmon pink, bright yellow; not go back to jolly, round-faced Father O'Riordan with his big, big bushy eyebrows, his deep Irish laugh, and his unexpected capers . . . .

The Easter Sunday morning when it was raining so hard . . . the parishioners came in with their umbrellas, shaking off their feet . . . and Father proceeds to walk out to the altar and begin to stride back and forth in front of the altar railing . . . his hand up to his eyes like an Indian scout searching the horizon in the distance . . . he looked first out the windows on one side of the church and then the windows on the other . . . finally turning to the congregation to say, " 'Tis a glorious morning!" And, of course, it was, for it was the Feast of the Resurrection . . . everyone howled; you just could not help it.

Not go back to the sisters with whom I had lived now for six years, just like one big happy family . . . .

The day we tricked Sister Mida about the fried chicken . . . it was Mother Lucian's day to cook . . . she had prepared mashed potatoes, gravy, and fried rabbit . . . we were not to tell the sisters who did not know that the dish was rabbit instead of chicken . . . Sister Mida looking all over the platter for her favorite piece, the chicken wing, and being unable to find it, finally settling on another piece . . . Mother announcing at the end of the meal that this had been rabbit and Sister Mida leaving what she had enjoyed as a delicious meal to go to be sick in the bathroom . . .

Not go back to the winters spent across the street from Mama and Sis who came down every year to improve their health in the good weather . . . .

Sis wearing an evening gown and passing out programs at the Spring Piano Concert; Mama ordering a beautiful cake frosted with piano notes to be served at the reception . . . just my knowing they were there across the street every day from January until after Easter.

All this now ended. I could not believe it! For though I had packed my trunk at the end of the school year as every sister in the Allegany Order was required to do, to make ready for a new assignment should one come, I had not for one moment entertained the notion that I would be transferred from Ft. Myers to some other location. I think, more than any other change I had made over the years, this one came as the greatest shock.

With my family, I spent the rest of the summer vacation reminiscing. There was so much to be grateful for and so many beautiful memories of Florida came to mind . . . .

The day I stepped off the Silver Comet train that had brought the other sisters and me nonstop from New York to Miami, I felt the warm, fresh air hit my body and saw the swaying of the fronds of all the different kinds of palm trees . . . the coconut palms with their trunks that bent every which way, growing wherever they bent; the royal palms that looked like straight poles of cement until your eye came to the bulbous, bright green top; the palmetto palms that almost looked like bushes; the travelers' palm whose leaf at its base formed a kind of cup you could drink from . . . oh, I took it all in. This was my place! A place of such beauty!

Mama had taken me to the doctor earlier that summer because I stayed so tired, often having to lie down on the floor by the piano when I was practicing, totally exhausted. This new assignment in this luscious climate was to be just the tonic that I needed.

The drive across the Tamiami Trail to get to Ft. Myers from Miami . . . my first glimpse of the handsome and inscrutable Seminole Indians . . . the women's beautiful, wildly colored skirts of bright strips sewed in different color combinations . . . the men's dark trousers and shirts as gay as the women's skirts, topped with a somber black hat that looked like a derby . . . men and women alike with braided hair and what to me looked like the saddest eyes, which somehow seemed incongruent with their brightly colored clothing. Peering deeply into the Everglades and occasionally seeing Seminoles' houses, thatched-roofed huts made of dry palms with four poles for corners and a high platform built for sleeping, at one point I was shocked to see a Singer sewing machine sitting on one of these platforms but quickly imagined one of the Seminole women sitting up there using the machine to sew together all those brightly colored strips for clothing. Alligators swarmed in the canals by the side of the road or sunned on the banks . . . and birds were so profuse . . . so many types of cranes and swamp birds. It was all so new to me. Even though I had been warned that this would be a very long drive, I had found it anything but boring.

Another night on the Tamiami Trail . . . Mama's and Sis's first visit to Florida . . . we drove over to Miami to get them . . . it was dark by the time we reached the Tamiami to return to Ft. Myers, dark and rain was falling in sheets. The road was no wider than two car widths and there were no shoulders. On both sides of the road lay deep canals filled with catfish, water moccasins, and alligators. We drove in silence and in fear, unable to see even inches in front of us. Then things got worse. The windshield wipers on the car stopped working! Nothing we could do—turning the switch off and on, reaching outside to push the blades—nothing would make the wipers start up again. So now we could not even see through the windshield. Yet we could not stop, for someone approaching from behind might hit us. I began to pray silently to my guardian angel. Suddenly we came upon the red taillight of a car in front of us. O, we rejoiced. We began to sing. For several hours we kept right behind that red light. The queer thing was that none of us could ever get a glimpse of the car itself; all we could see was just the red taillight. Then just as we were coming out of the Everglades, the red taillight disappeared. None of us ever did see that car! "But who ever saw a guardian angel with a taillight?" I chided myself, trying to understand what had happened. "Perhaps you have," was the only thing my second voice would answer.

That initial meal in the rectory when I tasted avocado for the first time! So delicious, served with a big blob of mayonnaise and salt and pepper.

The first slow southern drawl I heard . . . the fifth grade student who had come with her family to welcome us, a student who would later figure importantly in my life . . . walking toward us now with a big rag tied around her knee..."What in the world did you do, child?" I asked. "Ah fell down and Ah nigh 'bout killed mahself!" was her answer, except that with her drawl every word took on four or five extra syllables. I listened, enchanted.

My fascination with the strange floral and fauna. . . the cajaput tree whose bark you could peel back like a banana, right down to the middle of the tree . . . the guava trees with yellowish thick-skinned fruit that smelled like a cat . . . the orange groves (and the day we nuns were invited to go pick oranges in a neighbor's grove and came back with bushels and bushels, only to discover that we had misunderstood the directions and had picked the fruit of some stranger!) . . . the tamarind tree with a fruit like a big dark brown lima bean with a seed that you could suck, dark and gooey, like molasses.

The day we ran out of shampoo . . . Sister Loretta said, "Oh, my mother had gray in her hair, and sometimes the gray would look yellowish and she would rinse it with bluing. That's as good. Just wash your hair with some soap—your hair is whitish-blond anyway—and then use bluing." I took Sister Loretta's suggestion. Getting my supply from the laundry room, I filled the washbasin with water and put in a generous amount of bluing. Dipping down to rinse my hair, I heard Sister Conrad who was walking by exclaim: "What in the name of heavens are you doing?" "Oh, I'm rinsing my hair; how does it look?" "Oh, bright green," she answered. I said, "Yea, sure," and went on rinsing. But something in the tone of her voice made me wonder; I decided to squeeze my hair a little and lift my head to look into the mirror. What did I see? Blue streaming down my face, and a head of bright, kelly green hair. I finished rinsing as fast as I could, but my hair was still green. For weeks afterwards I went around with bright green hair. I could only be thankful that my head covering hid it.

The day Sister Loretta convinced me to give her driving lessons in the black four-door car someone had recently given to us sisters . . . "You've already been out driving," Sister Loretta said one morning when Sister Conrad was not available to take her out for her first lesson. "You can show me how." "Sure, I'll show you how," feeling like hot stuff after having been on the back roads in a sparsely populated subdivision a couple of times with Sister Conrad . . . Sister Loretta and I were now approaching a crossroads . . . "Do I turn right or left?" Sister Loretta asked. "Either way you want," I replied. "Just remember to put your foot on the brake to slow down before you turn." But Sister Loretta missed the brake and hit the accelerator. The next thing we knew we had jumped the curb, driven up through the middle of a lovely manicured lawn in front of a beautiful house, and run right smack dab into a big, gorgeous palm tree. The mistress of the house who was outside working in her garden looked up, startled at the sound of our collision, called to her child, "Get in that house immediately!," and then came to stand about twenty-five feet away from us, just staring. Two nuns in a black car smashed into her palm tree, my tall stove-pipe headpiece completely squashed into accordion pleats by my hitting the windshield. "Just back up. Back out of the tree and let's go," I said to Sister Loretta. "I can't; I can't," she cried. "You do it." So we shoved around, climbing over each other frantically until I was in the driver's seat. Then it hit me. "Sister Conrad never told me how to back up," I revealed to Sister Loretta. "This thing says reverse," *Sister Loretta answered, "try that." I did and, with the lady watching us all the time, we backed up and I drove around the block, out of sight, as quickly as possible. "Let's take a look and see what happened," I said to Sister Loretta. We got out, looked, and saw that the front of that car looked like the antlers on a deer—the front was pushed in all the way back to the radiator, from which we could now see water running. "We have got to get this fixed pronto," I told Sister Loretta. But as serious as the damage was, we couldn't help laughing when we told Mother Lucian.*

(The only thing funnier about the sisters learning to drive was the day Sister Mida was practicing in the field behind the school. Being instructed by us accomplished drivers to drive around the big oak tree, she just kept circling it, again and again. No amount of yelling at her would get her to stop driving in a circle. When we finally did get the car stopped, Sister Mida said, "I give up. Driving's just not for me." And we all agreed. None of us wanted to continue to try to teach her.)

The day I was accompanying the children who were being taught a minuet for a George Washington Day celebration . . . the visiting instructor wanted me to repeat a particular section of the song but I kept missing it. "Here," she said, "use this pencil to mark the place you are to come back to." So I made a big black mark on the sheet music so I would know where to come back to. Everything went fine thereafter— until the night of the performance.

The dance was progressing lovely; I had played the piece through several times, never missing the place to repeat because of the big black mark. About the fifth time I had repeated the passage, the big black mark got up and walked right off the paper. I had been looking at a fly instead of the pencil mark I had made during practice!

The day I rode a tree in the hurricane . . . It was a Saturday...there had been wind with a very little rain that had lasted for about an hour. Then there came an astonishingly loud roar, followed by heavier rain. When the rain subsided, we were watching out the back door when a tulip tree started leaning precariously. I could see the roots gradually pushing up out of the sandy ground. I felt so sorry for the tree; so in my innocence—perhaps ignorance is a better word for it—I opened the back door (with difficulty, of course) and, over the objections of the sisters, I made a mad dash for the trunk of that tree. Thrilling in the moment and feeling such a great sympathy for that tree, I clung to its trunk, slowly accompanying it on its journey. My arms stayed around the trunk of the tree as much to keep myself from being blown away as from affection for the tree, I suppose, but I felt as we went down that neither the tree nor I was afraid but were rather enjoying doing the unusual.

Yes, these were the memories that would not leave me that summer of '45 when I learned I was being transferred. This was the Florida I knew. This was the Florida that, no matter what, I knew I would never, never forget.

My family of sisters in Ft. Myers, Florida. I am first on the left.

This group represents a small part of the piano students I taught in Ft. Myers. Margaret Honc, now Sister Kathleen Francis (and the author of the introduction to this book), stands in the center in the dark dress. The concerts included pieces by Bach, Mozart, Brahms, Beethoven, Debussy, Chopin, and other greats.

Scott Hough, thirteen-year-old piano student of mine, who was frequently presented in concert.

This is Papa at the time he wrote the letters to me in New Orleans, 1951.

This collie is my very first pastel drawing, finished in one evening during summer school at Marywood College in Scranton, Pa, about 1946.

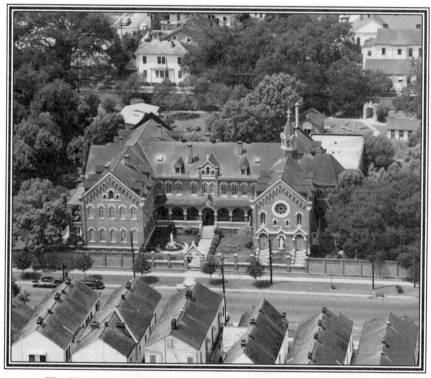

The Monastery of St. Clare, 720 Henry Clay, New Orleans, La.

In Those Days

Stretched along the mainland, separated by marshlands and bay waters from Atlantic City, are the cities of Absecon, Pleasantville, Northfield, Linwood, and Somers Point. Pleasantville was the center for shopping and St. Peter's Catholic Church. The other cities were mainly residential with few stores.

Northfield was the location for the Atlantic City Country Club. In those days the people in Atlantic City said you were "living in the sticks" if you lived in any of these communities. The "sticks" were hundreds of trees, many of which graced our property and adjoining properties.

It was a quiet and cool place to live. In those days nearly everyone in town knew everyone else. The car and the trolley were the main sources of transportation.

The trolley ran by only a block from our house. We often went to church by trolley and to school at St. Peter's. At the end of our block was an American store where we did most of our grocery shopping. No Super Markets in those days! You were waited on! They ground coffee for you and cut your choice of meat. No frozen foods nor instant coffee!

Nearby was a candy and an ice cream store. Popsicles were popular then, and for one cent you could buy a piece of candy that today would cost ten times that amount.

In those days, more than mail was delivered at your house. A horse-drawn wagon brought your milk with real cream at the top of the bottle. The ice man delivered a large block of ice every other day. If we were lucky to be there when he arrived, he would give us a piece of ice to suck on. Mr. Crowder came in his truck with choice pieces of meat. The vegetable man brought fresh vegetables. And if you heard a bell, it wasn't the ice cream man but the scissors' grinder to sharpen your knives and scissors.

Saturday was the big day in Pleasantville. In the afternoon, for ten cents, we would go to the Rialto Theatre to sit enthralled at the latest Tarzan picture and see Pathe News, Bugs Bunny and the serial Rin Tin Tin or the Green Hornet. But before that, we would stop in at the five and dime to buy some candy for the show. In the evening we might visit the brightly lit stores and certainly stop at the bakery for cakes.

We never visited the other communities much—Somers Point, Linwood, or Absecon. Absecon was the location for the exclusive Seaview Country Club. Al Smith, former governor of New York and Democratic candidate for President, often spent weekends there. He and his friends would go to Mass at St. Peter's in Pleasantville. We would often see him in his famous brown derby.

These communities were a good place to live, away from the hustle and bustle of any large city. The air was clean and cooler. The people were, for the most part, friendly. The only dark point was the presence nearby of the Ku Klux Klan. In fact, at one time they burnt a cross on our sidewalk. (Yes, we had sidewalks!) At another time, they threatened to burn down the wooden church in Pleasantville. Our grandfather, with other men, stayed up all night with shotguns prepared for any such threat. But for the most part, people of whatever religious or political persuasion, lived in harmony.

In the summer these communities were cooled by the shade of tree-lined streets and properties. And if you didn't go to the beach and ocean in nearby Atlantic City, you could always hop in a horse-drawn, two-wheel buggy and trot back to Bargaintown Lake. In the winter the trees were bare, but the streets were often covered with snow, affording us frequently the opportunity to use our sleds or have a snowball fight. One winter report read: "Snow today, followed by little children with sleds."

Times may have changed many things but not the memories of those quiet days of sled riding and snowball fights, of shaded streets, friendly people, a place where we were born, played and grew up in God's grace.

—George H. Muller, Jr.

Bruds, better known as George H. Muller, Jr., reminisced about the neighborhood during our childhood.

At the grate I, in white, become a Poor Clare novice and receive my new name, Sister Mary Bernadette. Chaplain Rev. Paul Callens, S. J., is assisting. To my right is Professed sister, Sister Mary Dominic.

I kneel in my white wedding dress at the grate during the ceremony of Reception.

I made this crown of thorns in preparation for becoming a Poor Clare nun.

This is the bridal dress I made for my reception of the habit.

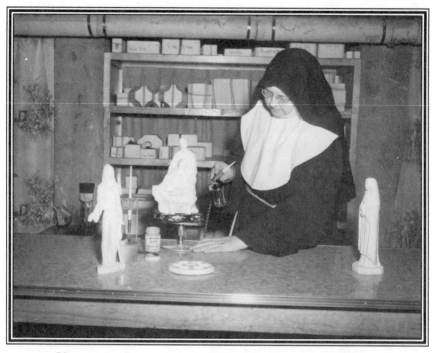

I learn to make ceramics at the monastery in New Orleans.

Yes, this is Papa, the mastermind and math scholar, with his great
invention—the great rake he built that he could hardly lift! Mama stands
on the right and Sis on the left.

SECTION FIVE

BOLIVAR, NEW YORK
ST. BONAVENTURE UNIVERSITY
HADDONFIELD, N. J.
AND
NIAGARA FALLS, NEW YORK
1946-1951

Everywhere Challenged by God

CHAPTER SEVEN

ENTERING THE "DARK NIGHT OF THE SOUL"

> With all my voice I cry to the Lord,
> with all my voice I entreat the Lord.
> I pour out my trouble before him;
> I tell him all my distress
> while my spirit faints within me.
> But you, O Lord, know my path.
>> FROM PSALM 42

So it was to Bolivar that I went.

Nestled among the hills, Bolivar was surrounded by the spires of pumping oil rigs. The tiny town was known at that time to be the wealthiest town in the United States in comparison with others of its same size and population. The town's name personified prestige, rank, status, and money.

The home that had been given as a convent seemed to me to be like a palace. Oriental rugs were everywhere; bedroom suites with adjoining baths that far outnumbered the sisters who would be assigned to that community. A downstairs drawing room held a grand piano, and a large front parlor had another piano. There was a lovely dining room that was seldom used and a sunny breakfast room off the kitchen. There was even a very lovely chapel with murals painted on the wall behind the altar, murals of St. John, the Christ Child, and Blessed Mother. The house was spacious and rich, and I felt called to "little and poor."

This house had belonged to Mother Mary of the Angels' parents. Upon their deaths it was bequeathed to the Allegany Franciscan Sisters. There were vague plans for the house to become a retreat for sisters who needed rest and recuperation, but at the moment it housed only the two of us. My assignment was as piano teacher at Bolivar Central School.

Every morning Mother Mary of the Angels and I would drive to St. Mary's for mass, and then after breakfast I would take the car on to school to give my morning lessons. The afternoons I devoted to my own practice, for I continued to be a dedicated student of the piano. In fact, Mother Jean Marie gave me permission to take the train into New York City every other week to study with Harold Spencer at Manhattanville College of the Sacred Heart. I would stay overnight with friends and return to Bolivar the next morning. I learned an amazing skill during those train rides into the city; I learned how to practice by merely looking at the music. I would imagine the intricacies in a particular piece, finger the technically difficult passages in my mind, and discover amazingly that some of these became very clear to me as a result of this unusual kind of practice.

Soon I began to have more students in Bolivar than I could handle, so I wrote to Margaret Honc down in Ft. Myers, Florida—the same little girl with the rag around her knee who had talked that slow southern drawl—and invited her to join Sister Mary of the Angels and me in Bolivar. Margaret was now a grown-up girl; she had often helped me in Ft. Myers by giving some of the younger students their piano lessons under my supervision.

"Margaret," I wrote, "Rev. Mother Jean Marie says that you can come up here if you want to, to help me teach these students music." I knew Margaret was seriously considering being a nun and this could be one way she could test her own intention. Margaret's brother Vincent brought her up from Ft. Myers and together we began to teach the students, often practicing ourselves in the game room in the basement the two-piano pieces we played in various concerts and recitals. It was wonderful to have Margaret with me there in Bolivar.

But all was not well. Living in this house reminded me so much of living at home with my parents. Our house had certainly not been so grand, for sure, but there had been the same sense of family present, the same intense interest in music, the same devotion to the Blessed Mother and to the Church, the same appreciation of culture and learning, the same love for color and smell and artful use of space that Mama and Papa had created both in our house and in our garden.

There were days that I would find myself standing for thirty, maybe even forty minutes, in front of the cupboard in the basement game room, staring at the large collection of dolls that Mother Mary of the Angels' parents had placed there. There were dolls they had collected in their travels all around the world, one which looked just like the Raggedy Ann Mama suggested I give to the poor little girl who lived in the gas station . . . .

"We're going up to Trenton Saturday to bring Sis from college for the Christmas holidays," Mama told me early in the week. "Why don't you take that Raggedy Ann you have upstairs on your bed—which you never play with—and give it to the little girl whose daddy runs that Shell station where we always stop?" Mama urged Bruds and me every Christmas to look for things we did not use or want that would be appreciated by other children. Well, while it was true that up until now I hadn't given the time of day to this Raggedy Ann, now that Mama was suggesting I give her away I had strong reservations.

Nevertheless, when we piled into the black Model-T on Saturday, I had the Raggedy Ann doll with me. But when we stopped at the Shell station for gas, I just could not part with the doll. Even when I saw the little poor girl, I could not give her my Raggedy Ann. On the way back, though, I had a change of heart. "Are we going to stop at the Shell station this time?" I asked Mama who always did the driving. "Do you want to?" she replied, checking, I suppose, to see what were my intentions. "Yes, I want to give the little girl my doll," I answered. So when we stopped, I left the doll with the child at the station, but I continued to think about Raggedy Ann and miss her. The doll had come to mean a lot to me after I began to consider giving her up. I just hadn't wanted to part with her.

I would stand for minutes at a time, too, in front of the big wall of pictures of Mother Mary of the Angels' parents—framed photographs taken in countries all over the world. One in particular fascinated . . . the parents' picture taken at the precise point that they crossed the equator. Was I thinking of my own Mama and Papa as I stood there? Papa, whom I had seen over the past many years only at vacation time when I went home for two weeks to New Jersey, his dedication to the job of city clerk of Northfield keeping him from coming with Mama and Sis each winter to Florida. Mama, who had become a regular part of my life as a result of the months she and Mary wintered each year across the street from St. Francis Xavier School in Ft. Myers. I knew that even right at this time the two of them were back down in their rented quarters in Florida.

The longer I lived in the house in Bolivar the more I began to lose touch with the simple things that had come to stand for my life in the convent. I began to lose sight of the spirit of true poverty, which is poverty of spirit. I knew the things in this house represented nothing in comparison with the love given us by Our Gracious Lord; yet, I found myself embracing them. I also realized that even with the peak of success I had reached in Florida in music teaching—all that, too, was ultimately empty. Wherever I looked now, I saw nothing but emptiness.

One day as I lifted up the piano bench to get out a piece of music, the truth of my situation hit me. I remembered another piano bench—this one lifted by a disadvantaged child about my age.

She was a child our neighbor had sponsored as part of a program to give children a break from the hot city ghettos by spending a few weeks in the country. Mrs. Rogers had told us that when this little girl got out of the car, she had stopped, looked at the grass intently, and then got down and rolled in it. She had never been in the country, so rolling in grass was a completely new experience. I had opened up the piano bench this particular morning when the child was at our house playing. The little girl had stood there looking at the things I had stored in the piano bench.

"Oh, my, you are very rich," she had said. I had looked into the piano bench, unable to figure out for the life of me why the child thought I was rich. This was just junk I kept in the piano bench. But I never forgot her statement. Even though we really weren't rich, we had so much more than she did. From that time on I had taken notice often of even the smallest things we had and felt the deepest appreciation.

Now, standing with my hand on the lid of another piano bench, this time in Bolivar, I was uncomfortable and actually entertained a sense of guilt. I knew one could reach sanctity and be holy in any walk and condition of life. But for me this place was a stumbling block. It was a situation God placed me in for reasons known only to Him. But I suspect that the dramatic situation that followed the raising of that piano bench in Bolivar was actually a great grace and the beginning of the "dark night of the soul" for me.

Each week my condition worsened. I began to be unable to sleep. I lost my appetite. Nothing around me mattered. I felt that I had to get away somewhere—somewhere out of this house— but I did not know where I should go. One day I reached the end.

Going to the telephone, I called Reverend Mother Jean Marie in Allegany. "I'm leaving here," I told her. "Where do you want me to go?" I could hear the startled surprise in Mother's voice, but she said, "My dear, you just come here to the convent. Come here to me, and we'll have a little talk."

It was a cold, cold snowy day that I fled from Bolivar to Allegany. Mother Mary of the Angels found a man who would drive me to the Motherhouse. He had an old truck without heat. My teeth chattered as I tried my best to make conversation to hide the terrible hurt and emptiness I felt in my heart.

"Reverend Mother Jean Marie wants you to have the room right near the chapel on the first floor," Mother Bernadine told me. As I approached the location, I thought, "How trusting. They're putting me right by the front door. Don't they know that every second I feel more and more like running?"

The next morning Reverend Mother Jean Marie asked to see me. "My dear, what is the matter?" she asked. All I could do was look at her and cry, "*Money, money, money!!* That's all I hear: *Money, money, money!*"

"I thought you were happy in Bolivar," Mother responded.

"I am," I answered. "I am happy there, but there's something that is the matter. I'm empty. That is just all I can say. I'm empty."

Reverend Mother Jean Marie was very patient and kind. "Now, look, dear, I'm going to let you stay here just as long as you want to. You can sleep in that little room down there and right next door is a parlor where there's a grand piano." Even though I had brought only one change of clothes, I had packed every single piece of my piano music. I practiced so much that the music seemed to be a part of me.

"Now, you just go in and practice the piano all you want; practice to your heart's content. Just don't be playing when the girls from the high school are in the chapel this weekend for their retreat with Father. But, otherwise, you may play all day long every day, if you wish."

I did exactly as Mother Jean Marie said. I practiced all day long every day on the grand piano in the parlor. When the weekend came, I determined the hours the girls would be in the chapel for their retreat; and then I refrained from practicing. Not wanting to be confined to my bedroom when the retreat was in session, I decided to slip in and sit in the back of the chapel. It was there that I first saw the priest, Father Juvenal, who was to figure so prominently in my future.

It seemed that every word that Father was speaking to the young girls applied to me; he seemed to be saying just what my soul needed. Yet, even with this inspiration, I still could not break out of my confusion. "What should I do? Where should I go?" The questions haunted me; nowhere could I see a solution. So when the retreat was over, there seemed nothing else for me to do except to go back to practicing my music.

It was while I was practicing the piano that I had the realization. The song I was practicing was "Clair de Lune." I can see the sheet music open in front of me yet. About half way through, suddenly, something seemed to strike me forcibly. I stopped playing, right in the middle of the piece. Something said to me— or something guided my thoughts:

"Sister Marie George, here you are a professed sister with vows and you can sit at this piano and work for hours at a stretch and you don't get tired. You don't get bored. But you, a professed sister, a spouse of God, cannot sit there in the chapel even half that long. In fact, you can't sit there half an hour without getting bored. *Something* is the matter."

I knew at that moment that there was a great imbalance. "All right," I said. "I will close this music and never touch another piano note again until I can go into that chapel and I can pray as long as I can play. For just as long a period of time as I am able to practice I should be able to sit in that chapel and not get bored.

But give up my music?

When music had been the ever-present thread binding together so many events in my life? Touching every place I had been and the people I had met? We had been a musical family who could have claimed to have been concert artists in the true sense, for the harmony of our music blended the art of living into beauty, peace, and satisfaction of accomplishment. It set moods for family living, was a means of expressing feelings of love, excitement, gaiety, patriotism, humor, beauty and even allowed release of pent-up emotions.

It's a cool day in the early spring. Mama is in the sun parlor—or the conservatory, as she liked to call it— practicing her music. Oh, how I loved to hear her practice. After she did her scales, she might warm up with "Nina," the song on the orthaphonic record Papa had surprised us all by bringing home one day, or perhaps she'd sing that aria from the operetta when she appeared as the leading lady on the Philadelphia stage shortly after she and Papa had gotten married. ("Your father and grandfather felt a music career and my duties at home would not be compatible," she'd explain to me when I would ask why she hadn't continued to sing in the opera.) I would sit there listening as she practiced, the sunlight pouring in the two sides of the room which were made of big windows, the ferns in the big white wicker baskets sometimes vibrating with the sound—oh, how I loved that one long wicker basket which stood in the middle of the room—oblong and on legs—with a big handle on it that went way up over your head. And hanging from that handle was the big cage in which Mama always kept her beautiful yellow canaries. I feel such a sense of security as I listen to Mama, a sense of inner happiness, peace, of being surrounded by love and a family sharing things. A wonderful "being together" feeling flooded over me.

And then the calm is suddenly replaced by excitement. From another part of the house, just at the moment Mama hits the high note finishing her song, we hear a sharp cracking sound. She and I run into the dining room to investigate, and there on the buffet lies the fine crystal bowl which always sat there—only now that bowl is in two pieces. Mama's voice had shattered that crystal! What pride the family took from that moment on in saying that Mama's voice was so clear and true that it could break crystal!

And music had come to be much more than an end unto itself. Rather, it was a part of me. Surrounded by its beauty from birth, unconsciously I had searched for it in my convent life. And it had been there for me, very much so. Music had become a means of expressing love for God. The hymns we sang really became prayer. The Gregorian Chant which I studied calmed the body and cleared the mind. Its purifying effect prepared the whole person in a way no other music could. This exposure to Gregorian Chant in childhood prepared me for the life style I was to adapt in later years. In the convent, I had not had to settle for less than the caliber of music I was exposed to in my early years.

In the climate of the convent, my interest in the piano had continued to grow. Realizing my attraction for the keyboard, my superiors had encouraged further study for me, and my early piano teaching was supervised. This gave me confidence which had led to the unusual piano teaching profession I was now enjoying. I say unusual because I have not, neither before nor since, met up with similar piano teaching circumstances anywhere as those I enjoyed in the Franciscan convents.

My music, too, had become for me a means of communication. A valse of Chopin, a Bach fugue or a song of Cadman managed to say something for me that words could never impart. Music also was now a means of my bringing enjoyment to others. Whether I improved the world of music was not important. I knew that through my teaching and my performing I was contributing to human happiness. And that was what counted.

No, I could not imagine life without music. And yet this was what I was now giving up. *Until I can sit in the chapel without getting bored as long as I can sit joyfully at the piano* . . . that was the demand I was now making of myself. So, the music went into the briefcase, and I went into the chapel.

There, I sat. And I sat. And I sat.

I came out of the chapel only when the bell rang to go down to eat. I came back, said the prayers with the community; then everybody would leave and I would sit and I would sit and I would sit. Waiting. For what I did not know. But I did know that no human being had an answer for me.

During the first days of sitting, I did something which is my custom never to do: I picked up someone else's book that had been left in the chapel on the pew right beside me. *Characters of the Passion* was its title. Impulsively, I picked up the book and opened it at random. One sentence stood out: "Purity is gained only on the knees in humility." "Well," I said to myself, "I need purity of intention in everything that I am doing. I need purity in everything." So I closed the book and I knelt. "From now on," I determined, "instead of sitting when I come into the chapel, I will kneel."

The days passed on. I did nothing now but wait. Wait before God, present there in that chapel in the tabernacle. For more days I knelt and I knelt and I knelt. Again, I stopped only to eat when the bell rang or to join the community when the bell rang for prayers. I can recall very vividly how hot my knees became; I could actually feel the heat radiating from them. And their color was as red as beets.

As time passed, I felt the need for some exercise, so I began taking little walks back and forth in front of the convent which was situated on a beautiful hill right above a highway. I could see the cars coming and going, and I would think: "Where are they going? Where are they coming from? Where are these people going?" I could still feel the heat from my knees radiating into the coolness of the morning as I walked. After these little walks, I would go back into the chapel and I would kneel.

I knelt and I knelt and I knelt.

I waited and I waited and I waited.

Without knowing just when it happened, very gradually I began to feel that I wanted to give God everything. I believed that in the past I had been holding something back from Him; but now I wanted Him to have everything. But how could I give Him everything? One morning as I knelt, the solution came to me. I would leave everything, and I would go to the cloister. The moment I made that commitment, I was overswept by a peace that was all-encompassing.

"I'd like to have an appointment with Reverend Mother Jean Marie," I told the sister who checked in on me daily. Upon my request, the nun's eyes flashed a look almost like fear and her head suddenly began to dart first in this and then in that direction. "What in the world is the matter?" I asked myself, completely puzzled by the sister's reaction. Then the awareness hit me. She thought I was going to announce that I was leaving the Order. She thought this was the reason I wanted to meet with the Reverend Mother.

Reverend Mother Jean Marie herself seemed concerned when I went into her room to meet with her. "My child," she said, her sweet smile coexisting with a look of perplexed uncertainty, "how are things with you? How are you?" I didn't even answer her questions before I blurted out, "Mother, I want to enter the cloister." Relief seemed suddenly to spread over her face, starting with the lines on her forehead and going all the way down past her trembling chin. I realized then that she, too, had been afraid I was going to ask for dispensation to leave the Order.

But she couldn't have been farther from the truth, for even in those darkest days in Bolivar when I walked from room to room almost smothered under a cloak of emptiness and blackness, I had never considered giving up my vocation.

So now as I faced Reverend Mother Jean Marie it was not to tell her I wanted to leave the Order but instead it was to await her reaction to my request:

"Reverend Mother, I want to enter the cloister."

"Oh, my dear, how beautiful," Reverend Jean Marie responded, words which were immediately followed by a question: "And now, my dear, are you willing to go back to Bolivar? Because I know you were planning a concert. They will miss you back there and will want you to come back. Will you go?" "Yes, Mother, I will go," I answered sincerely. Fidelity to my vow of obedience prompted this immediate response, for even as a child I had considered obedience to my mother the same as obedience to God.

"What I tell you to do," Mama had instructed us children from early on, "is God's will for you, coming from me. That's why you must always obey." I had accepted this teaching without question. "You know, I never have any trouble with Nina," Mama was saying to the neighbor when I walked up. "She asks me if she can do something and I say, 'no,' and she doesn't keep after me to try to get me to let her do it." . . . There was the time Mama was driving Bruds and me to Atlantic City. A car coming toward us around the bend did not take the curve but came straight at us so that Mama had to go off onto the shoulder of the road to escape being hit. Then we heard a crash. Mama had stopped the car and at the same time said to my brother and me sitting in the back seat, "Do not turn around to look. I am going to see if they need help." Then Mama left us in the car where, despite our intense curiosity, neither Bruds nor I even turned our heads a little bit to look. When Mama returned, she told us the steering gear on the lady's car had locked. She had run into the fence but was not injured.

When I thought of this now I marveled at how obedient we were as little children. We would no sooner have looked when Mama said not to than to fly to the moon.

This kind of training in obedience had prepared me, I suppose, for I had never had any difficulty obeying my superiors after I became a religious. Up until this day I have always been anxious to seize opportunities to obey in even the slightest things. It is a sure way to gain grace and to do God's will since He speaks to us through authority.

Although I felt she may not have taken my request seriously, I left Reverend Mother Jean Marie's room with a light heart, ready to do whatever was God's will for me. Now that I had told her of my desire to enter the cloister, my soul was at peace and I entertained thoughts that all would be well.

CHAPTER EIGHT

BEGINNING OF THE GRACES OF MYSTICAL PRAYER

> It is not ourselves we preach but Christ Jesus as Lord, and ourselves as your servants for Jesus' sake. For God who said, "Let light shine out of darkness," has shone in our hearts, that we in turn might make known the glory of God shining on the face of Christ. This treasure we possess in earthen vessels to make it clear that its surpassing power comes from God and not from us.
> —FROM SECOND CORINTHIANS 4

Back at Bolivar, I resumed my teaching duties as if nothing had ever happened. Margaret Honc and I presented a lovely and very successful concert on the stage at Bolivar Central School. Once again I was spending my days among luxurious Oriental carpets and glistening cut crystal chandeliers, living in palatial splendor.

But not for one moment, not for one day, did the thought of entering a cloistered monastery leave me. Every time I knelt to pray, I was practically overcome by the intense desire to show God how much I wanted to give everything to Him by leaving the outside world and entering into the cloister. Not one word came, however, from my Superiors. It was clear that they considered the experience of emptiness and its concluding outcome a spiritual episode which I would get over.

But as the weeks and months went on, the decision I had made was not something that I began to get over. To the contrary, the internal tension increased. I would not be happy until I could enter the cloister; that was where I felt God was leading me. Yet there was Reverend Mother Jean Marie's ongoing silence on the subject.

Then a bright idea hit me. Father Juvenal. The priest whose teaching I had enjoyed so much when, during the retreat of the high school girls at Saint Elizabeth's, I had slipped in and sat in the back of the chapel. If I could just get to him . . . all the things he had said that weekend had been so directly applicable to me . . . I had felt good hearing him . . . relieved from much of my heaviness. If I could study with him, perhaps he would be available for the spiritual direction which I felt I needed.

Reverend Mother Jean Marie wisely honored my request. "Yes, Sister Marie George, you may register for summer school at St. Bonaventure University to study with Father Juvenal." So come June Margaret Honc and I, she by now a postulant, moved to St. Elizabeth's for the summer and I walked every day down the hill and across the highway to attend the university.

When I wasn't in Father Juvenal's class in philosophical psychology or studying, I waited in the chapel just as I had done a few months before after I arrived at St. Elizabeth's in spiritual crisis. This was the place where I felt so much peace. I could hardly wait each day or evening to enter the beauty and quietness of the chapel. I had deep spiritual satisfaction in knowing that Christ Himself was present before me in the form of the Blessed Sacrament. I knelt there and I waited.

It was in the chapel one early evening that I had my first mystical experience.

A few sisters were scattered around the chapel, praying, and Sister Emerita was in the choir loft playing the organ. Occasionally other sisters would join her to practice some of the hymns which they would be singing for mass the next morning. At the moment I was praying for my sister. Mary's marriage was in difficulty and I was just putting that problem before the Lord in the Blessed Sacrament. During that prayer which was just from my heart, I experienced for the first time a Presence that even now is indescribable. The Presence of God within my soul. It was as if He were taking that prayer from me.

The profound experience of God's Presence made me oblivious to all surroundings. Even now I can recall vividly the physical feeling I had as the experience gradually faded away. I began to feel the pressure of my weight upon my knees. It was as if I had been lifted off my knees during that experience.

I suddenly wondered if anyone in the chapel had noticed anything unusual. I looked around the chapel quickly. Evidently, not. For the prayers and organ music and hymn practicing were going on as usual. "Perhaps there hadn't been anything to see," one voice inside me argued. "Perhaps nothing really happened." But another part of me knew differently. Something profound and ethereal had occurred in those past few minutes, something which had left me in extreme wonderment and joy.

I certainly couldn't talk with anyone about this strange and inexplicable experience. How was I to come to ever understand it? Ah, perhaps an explanation of what had happened to me lay in some book. I left the chapel and headed straight for the library.

Walking among the stacks, it was as if I were literally led to the shelf on which stood the life story of St. Teresa of Avila. My hand reached out for the book even before my brain and eye registered it. I opened at random:

On the Presence of God, felt in the "prayer of quiet": The soul understands, in a manner different from understanding by the exterior senses, that she is now placed near her God. . . . This does not happen because she sees Him with the eyes of the body or of the soul; for as holy Simeon saw the glorious little Infant only under the appearance of poverty, . . . he might have supposed He was the son of some mean person than the Son of the Heavenly Father. But the Child made Himself known to him, and so, in the same way, the soul understands He is there.

I could hardly believe this! I had been guided in just minutes to the exact source which could explain to me something of what had just happened—a book in the library by St. Teresa of Avila who was born in Spain in 1515! A nun who joined the Carmelite Order, who made great progress in the way of perfection and who enjoyed mystical revelations. St. Teresa went on to write books filled with sublime doctrine, books that were the fruit of her own spiritual life.

No, I was not addled nor confused, nor worse yet, off my rocker. I could now clearly recognize in my supernatural experience in the chapel an element of the Prayer of Quiet as described by St. Teresa.

I hurried across the campus, trying to catch up with Father Juvenal before he disappeared into the administration building. "Father," I sputtered, almost before I came abreast of the priest, "I think that I have experienced the Prayer of Quiet." Father Juvenal stopped in his tracks and stood staring straight at me. "Why, what makes you think that?" he asked. "Well, I've been reading the life of St. Teresa, and I think that what happened to her is the same thing that happened to me the other night when I was in the chapel." That is when he had blurted out to me, "Don't go looking for things like that. Don't go looking for those things. You can't . . . you can't *make* things like that happen."

"Oh, no, Father, I know I can't because I've tried. I wanted to repeat that experience I had of feeling God's Presence, but I didn't know how."

"Well, you let this be; don't concern yourself so much about it," was all Father Juvenal would say as we parted ways and he headed for his office while I walked back up the hill to St. Elizabeth's Convent. But as truncated as the conversation had been, I felt good about having had it. Somehow, I felt that Father Juvenal knew what I was talking about, and it felt safe and reassuring to have informed him of what had taken place.

But my unusual experiences during prayer were far from over. A few days later I joined the professed sisters who lived permanently in the St. Elizabeth's Convent when they went into the chapel after dinner to pray the fifteen decades of the rosary. This was a lengthy prayer: the Joyful Mysteries which consisted of five decades of the rosary; the Sorrowful Mysteries made up of a second five decades of the rosary, and finally the Glorious Mysteries which also included five full decades of the rosary. The sisters had a set pattern they followed when they prayed the three mysteries: the first five decades were prayed kneeling, the second five sitting, and the third five once again kneeling.

We began the prayer this evening, then, on our knees, praying the ten Hail Marys which made up each decade—"Hail, Mary, full of grace; the Lord is with thee. Blessed art thou among women and blessed is the fruit of thy womb, Jesus." At the precise point during the first Hail Mary when we reached the words, "fruit of thy womb,"—not one word before nor one word afterward—I experienced a joy, a presence that was beyond description.

Much later I was able to express a comparison: "It was," I told someone, "as if my soul were a beach over which the Divine waves were breaking joyously then gradually, gently ebbing out." I have often wondered how this process could have taken place during the short time it took to say those four words. It seems like an eternity happening, although the amount of time it took the sisters to say "fruit of thy womb" could only have been a few seconds. But there was no time for me in what was happening. Time meant nothing. It was impossible even to think of this occurrence in a framework of minutes and seconds, for it was truly timeless.

By the time this experience had repeated itself fifty times— ten Hail Marys in each of the five decades—I was afraid to move up on the seat for the next decades of the Sorrowful Mysteries. "What if the experience stops happening?" I wondered to myself. "If I move, that may be the end of experiencing this grace." "But if you do not sit down," another voice in my head argued, "you will stick out like a sore thumb when all the other sisters are seated."

So at the appointed time, I removed myself from the kneeler, took my place on the pew like all the other sisters, and we began the second set of Hail Marys. What would happen when we reached the words "fruit of thy womb"? Would I once again feel that unexplainable heavenly flooding?

I need not have worried, for the moment we reached that place in the decade, I had exactly the same experience. And in each repetition thereafter. Fifty more times I felt waves of joy at exactly the same spot in the rosary. We knelt again for the closing Glorious Mysteries. I knelt like all the others. Again, moving, changing positions, did not matter. Fifty more times, exactly at the words *fruit of thy womb* I felt the flow and ebb of a Presence. I could hardly fathom what had happened as I rose to leave the chapel, but I knew that after this experience I would always have a great love and appreciation for the rosary. And I would always have a deeper love for the Blessed Mother.

The next day after class I once again ran after Father Juvenal, this time barely catching him before he went into his office building. I blurted out, "Here's your pest, Father. I'm like a mosquito bothering you. But it happened again. Something amazing happened to me in the chapel."

Once again, I told Father my story. Once again he remonstrated. "Listen, Sister Marie George, when you feel something like this coming on again, just go do something else. Don't let yourself be available."

"But, Father," I questioned, "how then do I pray? What *is* the right way to pray?"

"Just say the Lord's Prayer," Father replied. "And say it slowly."

"But . . ." I inserted.

Father interrupted, "This is enough for you. This is how Christ told us to pray. So you should be satisfied with that. Just say 'Our Father.'"

That evening after I attended a concert presented in the convent auditorium, I walked down the stairs to the choir loft above the chapel. "I'll just kneel here," I said to myself, "and say good-night to Jesus in the tabernacle." As I was kneeling there, I began to experience His presence ever so slightly. "No, Jesus, Father Juvenal said I should not be looking for this," I spoke aloud, getting up as quickly as I could and leaving the chapel choir loft.

I went directly to the dormitory in which about twenty of us slept in beds partitioned by curtains. I donned an old night habit which I had been wearing lately and hopped into bed. "Now," I thought, "I will pray as Father Juvenal has told me to pray!" On the flat of my back, I commenced very slowly, "Our Father, who art in heaven, hallowed be Thy Name." I shall never forget what happened on the next three words, *Thy Kingdom Come*. The simplest way to put it would be to say "It did! The Kingdom did come!" All purity, all beauty was suddenly there within. Never had I experienced such love. Sensing the Divine so close I could not understand how such ALL Goodness could abide within my sinfulness, and I immediately attempted to say inwardly, "Oh, how can this be, do You not remember. . . ?"

I made an effort to name a particular sin which I had committed but was absolutely unable to recall it myself; and, instead of the word coming to mind, in its place there flashed an intense Light so bright it would have paled the sun. This was the Light of lights. My soul cried out in wonder and I asked, "Is it the Holy Ghost?" This I asked because there was such an experience of Love. I had been taught that the Holy Ghost is the Love spirating between the Father and the Son.

In answer to my question there was silence; nothing said it was and nothing said it was not.

I now imagine that my next question was asked because the Son was familiar to me. "Is it the Son?" Again that perfect silence, not saying it was or it was not.

There was only one other possibility which seemed to me unlikely, for I had looked upon the Father as I had seen Him portrayed by so many artists, as an old man with a beard sitting on a throne, unapproachable. Not a very intelligent perception I admit, but mine, nevertheless.

"Is it the —?" And here again in place of the word, all I can say is "It was!" Simultaneous with the thought *Father*, He was All in All, overpowering Beauty, Goodness, Purity, and All things good!

I was led to repeat my questions in the same order, over and over again through the night; and each time the experience of God was repeated. At the onset of this Presence I discovered I was absolutely unable to move my arm or any part of my body. Consciousness of my body had faded away and I was entirely wrapped in Him Who possessed me. Time ceased to exist. Eventually the experience began slowly to fade away but it was awhile before my body seemed to respond. Never had I experienced anything like this. I was not even aware such a wondrous grace existed in this world!

All the rest of the night I lay in peace and wonder. I could hardly wait for morning when it was time to get up. I felt like going up on the roof of St. Elizabeth's Convent and shouting to the world, to the cars going by, to everybody, "Oh, this is how it is!" I had such joy and yet I was torn apart inside, knowing that there were no words to fit, no mind (including my own) to fathom, the love which awaits us all! Never before had I been so anxious for morning to come and time for Holy Mass.

The chapel was full—all the visiting sisters going to St. Bonaventure University for summer school, all the novitiate sisters and the permanent residents of St. Elizabeth's. We sat in chapel according to the number we had received when we came into the convent. My number was 726. I went to my seat, hardly able to wait until the bell rang announcing the priest's arrival. When the bell finally rang, something jumped, leapt inside me. Then I was hardly able to wait for Holy Communion. As we stood to go forward for communion I felt as though I were running. Father Celsus Wheeler offered that mass and the moment he placed the Sacred Host on my tongue my whole being experienced a completeness indescribable. Much later, remembering this incident, it seemed that receiving the Son made my experience of the Trinity complete.

For the rest of my life the memory of that night and that morning will remain with me as vividly as if the events had just happened. In fact, a year later, when I was teaching in Niagara Falls, one day I sat down at a typewriter and slowly pecked out the following verses, never to this day changing a word. The verses tell the story. They are, to me, a form of prayer, lifting my mind and heart to God. As a result of this grace given so unexpectedly, my whole spiritual life has taken shape. That experience has affected everything I do. Although I know that "Eye has not seen, nor ear heard, what things God has prepared for those who love Him," I know I have had a taste of heaven, and I want to go there where He is.

"THY KINGDOM COME"

Thou camest to me
Camest from me
From within me.

I

Beneath all purity of Thine
This soul was speechless
'Mid Word Divine.

II

'Neath God of goodness
Unfathomable Wonder
This soul to press.

III

Soul perplexed by All in All
Puzzled by such condescension
Strived the ugly to recall.

IV

Tried so hard, God, to remind
O blinding Light!
No sin to find.

V

Could it be the Holy Ghost?
The soul awaited
Answered not this Holy Host.

VI

Spoke the soul of God the Son
Then awaited, yet,
Answered not the Holy One.

VII

With swift assurance and delight
'Tis God the Father
Of might, of Light!

VIII

The soul then spoke again, again
(Led on by Love)
Those sacred Names.

IX

Though no earthly word was spoken
His Presence surely known
Heavenly Manna, Mystic Token.

X

Loving Love, the soul exchanging
Carnal for the spirit's delight
Secrets seemingly unending.

XI

'Tis true these eyes have never seen
Nor ears have ever heard
Soul's secret, God entered in.

XII

All her faculties suspended
One with God, her
Cares all ended.

XIII

Rays of light the soul caress
This light, the True Light
All other light mere darkness.

XIV

No longer just another guest
United, Father and Holy Spirit
Within soul's center lies in rest.

XV

True Completeness incomplete
Borne along and lifted up
Borne along the Son to meet.

XVI

The soul received the Holy One
Then Trinity complete
Father, Holy Ghost, and Son.

XVII
O fire of Love, O Fire Divine
All creatures share
This love of mine!

But Father Juvenal had given me strict instructions not to dwell upon my prayer experiences and when thoughts of entering the cloister came to mind, to banish them immediately. "Confine your reading to the New Testament," Father instructed me. And that I set out to do. But one thing the priest had said puzzled me very much. "What you are experiencing is not extraordinary," he told me. At the time I could think of nothing to reply, but I know now what I was thinking but never said: "The experiences may not be extraordinary for you but they certainly are for *me!*"

And I wondered why all these years if such were ordinary I had never heard anyone relate such experiences of God. (Later, of course, when I understood more about the infused graces, I would come to recognize the wisdom of Father Juvenal in testing my experiences and in directing me not to put undue emphasis on external signs but to concentrate on the contriteness of the heart.)

The longing to give myself to God intensified day by day as I strove to do my best in a new assignment. I was sent from Bolivar to Haddonfield, New Jersey, again to teach piano. The change could not have been more dramatic. Although the convent in Haddonfield was a family home, it was extremely confining and not a little uninviting. However, the sisters with whom I lived made it a pleasant situation.

I especially enjoyed one particular assignment. "In addition to teaching children piano, you are to play the organ for the junior and senior choirs at church," Reverend Mother Jean Marie informed me. This I found exciting. The organ was a big Hammond that gave forth boisterous sounds; for the first time I even had chimes available to me, and I loved playing these at Christmas. It was here at this parish church where I played the organ for America's first televised mass.

I was also given another job, one I found curious: I was to count the money taken up in the Sunday collection. (This reminded me of the crazy job I had early on after becoming a sister. Mother Solanus in Rochester would not hand a dirty bill to anyone. So each week it was my job to wash the paper money the convent used. (If you want to see crisp new bills, wash them with soap and water and iron them flat!) But now in Haddonfield the bills, of course, were not hard to count; it was the change that gave me all the trouble. I finally caught on to how to use the little money-wrapping machine, and I'd spend hours every Sunday afternoon dropping the change into the little funnels at the top of the machine, letting it trickle down in its respective denominations and then fall into the round tubular papers which I then folded and stacked.

I also realized while at Haddonfield what little concept I had of how to manage money. One day the sister who cooked got the mumps so someone else had to go out to do the shopping. I was tapped for the task, and this was the first time in my life that I had even been into a grocery store to buy a large quantity of groceries. When I gathered all the items on the list and checked out at the register, the young clerk exclaimed, "Oh, oh, wait until sister sees this bill; her bill is never this high!" "Well, I just bought what she told me," I said to the girl. "But you didn't think about brands, did you" she responded, "or comparison shop to get the best price on the item?" No, I certainly hadn't done that; I didn't even know that was how you should do your shopping!

Here in Haddonfield I no longer enjoyed the experience of God's Presence but faith told me He was close and I tried to please Him by performing my ordinary tasks well. There were ways in which I knew He was looking after me. One day I was down in the basement doing the laundry, using a washing machine with an electric wringer. While putting the wet clothes through the wringer, using my hand to guide them, somehow or other my fingertips got caught in the wringer. Before I could do anything, my whole hand went through the wringer! The wringer quickly jammed but not before the damage was done. My hand was completely on the other side of the wringer.

"What are you going to do?" I exclaimed to myself. "How are you going to get this hand out of the wringer?" The only thing I could do was to turn the wringer backwards, but this meant my hand had to come all the way back through it. "You'll just have to endure the pain," I told myself. "There is nothing else you can do." So, closing my eyes, I turned on the wringer.

Oh, my hand hurt terribly. And when I let myself really look at it, I could hardly believe what I saw. Instead of a normal hand with a back and front, my hand was now completely flat! "I have to do something about this right now," I cried, hurrying toward the sink to put my hand under running water. "I have to play the organ tomorrow and I'll never be able to do it with this flat thing!" I began to say Hail Marys. In just seconds I felt a strange sensation in my hand and the pain stopped abruptly; when I opened my eyes and looked, my hand was no longer flat. It was perfectly normal! I marveled about this for days, for not only did my flat hand return in seconds to normal, but I never had a single ill effect thereafter. No bruise; no pain; no aching. To me this was just amazing.

I was grateful for having been stationed so near my family in New Jersey. It was while I was at Haddonfield that I first recognized Mama's physical decline. She, Papa, Bruds, and Sis came up to visit me before Mama and Sis headed back down to Florida for the winter; this was the first time I ever saw Mama with a cane. I knew she was failing. The presence of this new frailty, which, of course, I could not help but notice, deeply saddened me.

———————

At the end of the school year I once again returned to St. Elizabeth's Convent and classes with Father Juvenal at St. Bonaventure University. By now Father was president of the university, but he still took an interest in me. Calling me into his office one day, he asked, "Did you take those walks every day this past year, as I instructed you?" "Yes, Father, I did," I responded. "One of the sisters and I took a long walk every day for blocks and blocks, whether snow was on the ground or rain was falling." "Good," he answered. "That is very good."

"And what about that desire to go into the cloister? Do you still think about that?" Father Juvenal continued. "You told me not to dwell upon that," I answered. "And I have obeyed you. But, in all honesty, the thought is there in the back of my head at all times. So now when you ask me, yes, I still want to go."

"Well, now I will tell you that if you want to go, you will have to be the one to do the writing. You find a place if you have that desire."

"Will you give me a recommendation, Father?"

"You find the place first," was the only thing he would answer.

So that is how, over the next two years, I wrote letter after letter, being turned down by monastery after monastery. Those were hard years; not only was I having no success being accepted into a cloistered monastery, but many other things contributed to my gloom and unhappiness. Mama died during that time—a blow more terrible than I could even fathom.

Niagara Falls, the assignment to which I had been transferred after Haddonfield, depressed me. The moment I saw the place—the physical environment inside the convent cramped, dark, and dingy; no yard, only the kind of urban sprawl you find in the inner city—I felt a terrible, sinking feeling. Everything in me just went zoom, right down to the bottom of my shoes. On top of everything, I was not assigned to teach piano here but seventh grade, and I had no idea what to do with those big strapping Italian boys who seemed to fill up my classroom.

So in Niagara Falls I never sat down at a piano. After closing that sheet music of "Clair de Lune" that memorable morning at St. Elizabeth's I had never touched the piano again unless I was asked to or required to by my teaching. But until now working with my students had continued to keep me connected with the piano and my music.

Reverend Mother Jean Marie had asked me the summer before to write out all my music qualifications. "Make it sound good," she told me. It seemed the Order was contemplating making an office for a music director, a person who would travel from convent to convent, encouraging the nuns who were teaching music, showing them things they could do with their students. I had been tempted—I was thinking of all the wonderful things I had done with my students back in Florida—but even though I knew I would probably like the job, I never ceased wanting to go to the cloister.

I almost stopped eating during this time in Niagara Falls. I did not have time to eat breakfast since I played the organ for several masses before rushing over to teach the seventh grade. And at dinner it always seemed to me that there was not enough food, so I would refrain from taking a potato or piece of meat, hoping to leave enough for the sisters at the end of the table. Usually, however, food would be left over, so the next time the cook would cut the portions and there really would not be enough for all of us to eat. I was hungry most of the time and lost a lot of weight. In fact, by the time I left Niagara Falls I weighed only 114 pounds, much too little for my frame.

But then came the blessed day.

The letter came!

After applying at least seven times to different monasteries—including three times to the sisters in Memphis whom I also tried to bribe by sending them all the candy my seventh grade students had given me at Christmas—(leaving out, of course, Philadelphia which had no yard and was right in the middle of the city and where they did not cook their eggs well done at breakfast)—I was now going into the cloister. One of Father Juvenal's friends, Father Michael Harding, had suggested to me after Father Juvenal talked to him that I write to the New Orleans Poor Clares. I had done so, and now here on the floor of my room at the head of the stairs lay the precious letter. I opened it and read the news. I had been accepted.

SECTION SIX

NEW ORLEANS, LOUISIANA
1959-1963

Another Major Move in Sight

CHAPTER NINE

THE NATURAL AND THE SUPERNATURAL GO HAND-IN-HAND

> While we live we are constantly being delivered to death for Jesus' sake, so that the life of Jesus may be revealed in our mortal flesh. Death is at work in us, but life in you. We have that spirit of faith of which the Scripture says, "Because I believed, I spoke out." We believe and so we speak, knowing that he who raised up the Lord Jesus will raise us up along with Jesus and place both us and you in his presence.
> —FROM SECOND CORINTHIANS 4

It was while I was busy with the bookbinding work in the library of the cloister in New Orleans that I found it. The book that was to make a great impact on my spiritual life. It was a small volume dated 1930. *The Kingdom of God in the Soul* it was called, written by Father John Evangelist. The little book was one in a series of Capuchin Classics (the others we did not have) written first in 1620 and reedited and published in English in 1929.

The Kingdom of God in the Soul became a treasure for me. I poured over its contents. Chapter 1: "Of the Great Ignorance of Men in Searching for Their Happy and Blissful End Which is God" . . . another: "That the Soul's Difficulties in This Search Proceed Wholly from Herself" . . . another: "The Manner of the Way the Soul Must Walk to God" . . . and still another: "Three Signs by Which the Soul May Know He Is in the Right Way to God" . . . and this one: "Of the Necessary Preparation to Find the Kingdom of God in the Soul." There were thirteen chapters in all, with so many, many wise things said in them.

I couldn't have found this little book at a more opportune time; I was in great need of encouragement and that tiny volume provided spiritual direction. Once more I was beginning to experience graces of infused contemplation similar to those I had experienced that summer when I was studying at St. Bonaventure University with Father Juvenal. Although each grace brought with it great peace, I longed to learn more about these visitations. Father Juvenal, of course, was no longer available to help me know how to think about this realm of mystery. The little book from the monastery library, therefore, brought both direction and consolation to my soul.

Lying on my narrow bed one night a few weeks after I had begun studying *The Kingdom of God in the Soul,* I found myself wrapped in prayer. I was thanking God for all the past graces in my life. I was thinking about all the times I had abused so many graces God had given me and yet how He continued to shower upon me so much love in spite of my many failings. So many times when I had started down the wrong path He had put me back on the right track. Then I thought of so many souls who couldn't seem to put their lives together. God is so lavish with His gifts and then we turn around and abuse them.

These were my thoughts as I lay flat on my back that night, praying. Suddenly it was as if God pulled a question out of me: "God, you made the gift of sex so intense and so attractive—what joy on earth could possibly surpass it? Why have You given such a beautiful gift to humankind when there was such a chance that it would become, when misused, such a sorrow to You?" Then into my mind came a rush of memories. . . .

It was the day before I left home the first time to enter the Order of the Franciscan Sisters of Allegany. Papa had called me into his room. "Do you realize," he said, speaking very seriously, "what you are giving up? A home, children, and freedom to do your will?" I, only sixteen years old, assured Papa that I realized all this and that it was my will now to enter the convent.

I am sitting with the other sisters in Wood-Ridge, New Jersey. The sun is shining in through the beautiful stained glass window, striking the children sitting in front of us at the mass, causing them to look almost angelic. One tot is dressed in velvet and wearing little black shoes with a strap across the instep.

I am three years old again, back in Pleasantville. Mama is holding tightly to my hand as we walk along a street. I am so tiny that all I can see is people's feet and occasionally my reflection in one of the big plate glass windows. Suddenly, into my view come the feet of another little girl, maybe a little bit bigger than I. But this little girl is not wearing the kind of shoes Mama always made me wear—ugly brown high-top shoes someone else always had to button for you! "Good for your ankles," Mama would always say, when I complained that I didn't like them. No, this little girl coming toward me wore another kind of shoe: beautiful little shiny black slippers with a strap across the ankle. Oh, how I longed for slippers like that. I, too, wanted shoes that showed your feet, shoes that were pretty. How early I had been drawn to things that were sensual.

I'm back again in the chapel in Wood-Ridge, New Jersey. The beams of sunlight shining on the children seated in the chapel drew my eye to the beautiful blond hair of a young child seated directly in front of me. Before I even realized what I was doing, I found myself reaching out to stroke the lovely soft strands which looked like transparent gold the way the light was falling on them.

I am a teenager, fifteen years old, wearing my first silk stockings—you could get them for ten cents a pair at Woolworth's Five and Dime—and a tiara in my hair. "Take that thing off your head," Mama pouted every morning. "A tiara is not something you wear to school; a tiara is for evening." "But it holds my hair back," I argued, "and makes getting ready so easy in the morning." So, in spite of Mama's protests, off I went to school every day wearing my tiara.

"What kind of bleach do you use on your hair?"

"How do you get your hair so white?"

"What do you put on your hair?"

I noticed that for days different girls from the senior class had been coming up to me and asking these silly questions.

"I never bother to put anything on my hair," I told them. "I don't need to. It is just naturally this color."

In spite of all this questioning, I was completely surprised by what happened. When The Spotlight, *our school newspaper, ran its special issue listing the senior class's choices of celebrities, there I—a lowly sophomore—was among them! Girl With Most Beautiful Hair, the caption read. And underneath, my name: Nina Muller.*

I am still in high school. We have just finished physical education class. Suddenly, the tenth grade gym teacher storms into the locker room, yelling: "If there is any girl in here that is just too modest to get undressed the way she is supposed to, then that girl should not be coming to this school. The showers are for taking showers; they are not some girl's private dressing room!" All the while she was hollering, I remained crouched in the shower where I ran every day after physical education to get dressed. "Just stay here quietly," I admonished myself. "She'll finally leave," all the while that I expected her any minute to cross the room and yank me out of the shower. In spite of her angry pronouncement, I could never make myself change clothes in front of the other girls. "It just isn't right," I told myself. "It's God's will that we be modest."

Every day at lunch the school band would play and a big bunch of us girls would start dancing. Oh, how I loved to dance! Mama couldn't believe it when I wore out a brand new pair of shoes in less than a month from so much dancing. (Once for some reason the school band stopped playing at lunch and I promptly went to principal: "How come we can't dance anymore?" I challenged him. After we talked a while, he gave me permission to go find the band members and tell them they could start playing.)

Since we only danced girls with girls, I looked among my classmates to find a partner as soon as the band started playing. The girl I chose was not any too pretty, but she was an excellent dancer. Well, my friend and I were dancing away when we are approached by this handsome fellow, a senior. When he got to where we were, I heard him say, "May I dance with you?" and since I thought he meant the other girl, I answered "Oh, sure," and turned to leave the dance floor. "No," he said, "I mean you," reaching for my hand. In a moment I was moving around the floor with the first boy I had ever danced with, and did I enjoy it.

It wasn't until a few days later that I realized we must have made quite a spectacle. When Mama or Sis and I danced, we always held each other real tight; I would lean against Mama or put my head on Sis's shoulder or against her cheek. I assumed you danced this way with any partner—it was the way to dance, it seemed to me—so I had held on tightly to my male companion and put my head against his cheek. But was I surprised when The Spotlight *carried my name for the second time! "Bob Albrecht...who is this he's dating now? This beautiful blonde Nina Muller?" and all that stuff. I thought it was great, although he certainly wasn't dating me. One dance and that was it. But I loved the attention.*

"You also really enjoyed sitting on that same boy's lap," the adult voice inside my head spoke critically. "Well, yes, I did," I allowed, "but you know that was totally innocent. With six of us kids piling into one car after dancing around the bonfire at the street dance, there wasn't any place to sit . . . and I certainly didn't plan to end up in the lap of Bob Albrecht!" "But you were thrilled when it happened," the critic spoke back. "Yes, I admit that was a big deal for me, but you know that nothing ever developed between us."

By now, I was feeling very defensive. "You know," I continued to argue with myself, "that if there ever did surface a conflict between my relationship with boys and my religion, there was no question that I always chose my religion. What about the night Pete Starn took me dancing?

Mama greatly approved of Pete Starn. "Pete's a nice boy," she would say, and when he sent me a corsage to wear to his high school dance, she added, "And he knows how to do things properly. Go look in the icebox and see what's there." I did, and in a lovely box resting in fluffs of torn paper lay a beautiful gardenia.

"You can wear my pink organdy dress and my picture hat," Sis had told me. I felt like a lady when Pete came to pick me up; he was such a gentleman and escorted me so gracefully. But when we took our seats on the trolley car to go over to Atlantic City to the dance—Pete attended Atlantic City High School while I was at Pleasantville—I really got embarrassed. My date insisted on leaning over from the

*right-hand side to smell the gardenia on my left shoulder. I
was greatly embarrassed.*

*We danced until almost twelve and then went for a walk on
the boardwalk. "I'd like to take you to get something to eat
before we go home," Pete said. "Oh, but I can't," I said,
looking at my watch. "It's after midnight and I always go to
mass on Sunday morning; we always fast on Saturday night
from midnight on if we planned to take communion the next
morning." So I couldn't go eat; that was just out of the
question. Pete wasn't a Catholic, so I tried to explain the
rules and regulations; and, although he was nice about it, it
was clear that he was aggravated. Well, home we came, even
if it did mean cutting short the party.*

As if in answer to all this questioning about the attractions of
sexuality, I was suddenly caught up in an experience of God
which I find very difficult to describe. The best I have ever found
was someone else's words—the words of the Venerable Mary of
the Incarnation: "The Divine Word, taking possession of my soul
and embracing her with an inexplicable love, deigned to unite
Himself with her and to take her for His spouse. When I say that
He embraced her, it was not after the manner of human em-
braces, for nothing of that which falls within the cognizance of
the senses in any way approaches this divine operation; but we
have to express ourselves according to our gross way of speaking,
since we are composed of matter. It was by Divine *touches* and by
penetrations of Him in me and of me in Him."

All I can add is that I was conscious of Divine movement in
every single cell of my body and soul. Never before nor since has
that happened to me. It left me with an experiential knowledge
which could never have been realized if God had not given that
particular grace. All earthly joy, pleasure, happiness, and delight
became as nothing in comparison with this experience of God. It
was a foretaste of what awaits us when we leave this earth.

Another mystical moment occurred around this same time
and was also very special to me, creating peace and calmness in
my soul. On another evening, as I went to rest, I became suddenly aware of His Presence and was filled with great joy. Then,
somehow, I became aware of the presence of my mother who
passed away in 1949. Then I heard the Divine Presence say, "Ask

and you shall receive." Somehow I rapidly flashed back, "That Papa may be where we are." And then I immediately heard my father's voice, "It's been one whole year." Together we were lost in God. The experience gradually faded away. When I came out of the state of prayer, I thought, "How typical Papa's remark was since he had been a mathematician." Then and only then did I realize that this was the anniversary of my father's death. He had been dead one whole year! Through the experience of this grace I felt that Papa had been released from purgatory and was now in heaven with Mama.

CHAPTER TEN

A CUBAN "ROSE" ENTERS MY LIFE BUT NOT WITHOUT THORNS

> May the Lord make you overflow with love
> for one another and for all,
> even as our love does for you.
> May he strengthen your hearts,
> making them blameless and holy
> before our God and Father
> at the coming of our Lord Jesus
> with all his holy ones.
> —FROM 2 THESSALONIANS 3

Something was about to happen here in New Orleans which would change my life every bit as drastically as had the experiences that led me to come to the cloistered monastery, something which would change the total structure of my days and nights and alter my very life's direction—the arrival in 1961 of the Cuban sisters.

For weeks Mother Margaret Mary had been acting secretively. Perhaps we'd be at supper and she'd be summoned to the telephone; we could tell from the way she jumped up from the bench and rushed toward the office that this was an emergency. Yet when she returned to finish eating, she would never even allude to the conversation. One day I came upon Reverend Mother as she was first measuring the width of the hallway and then going into one of the sister's rooms to measure the bed. I could not imagine for the life of me what she was up to, and she offered no explanation. And Sister Mary Roch began to make many unexplained trips into town . . . we heard snippets of conversation about extra household supplies and extra food—and even once something about a secret telegram!

Then one day Mother Margaret Mary summoned us to hear the news. The call spread rapidly through the monastery. "Reverend Mother wants everyone in the library right away; she has something very important to tell us." All thirty-six sisters crowded in; we were all full of anticipation and some of us were even downright excited. Something strange was afoot and now we were about to find out what was happening.

Mother Margaret Mary began her speech very somberly:

"I know you are all aware that a revolution is taking place in Cuba, for we have been praying daily for our Cuban sisters. But what you do not know is that the Poor Clares in Havana are in great danger. The fighting in the streets has spread. Revolutionary soldiers have entered and taken over the grounds and buildings of the monastery. Historical records, business accounts, prayer books, archival materials—all have been heaped in the middle of the courtyard and burned in a massive bonfire. Sacred relics and precious religious statues are no longer safe. Huge numbers of soldiers bivouac in the monastery every night, some of them even sleeping in the sanctuary of the chapel. And I have just been informed by a friend of the Cuban sisters who was able to slip onto the grounds and then slip out again that two nights ago the monastery gardener was murdered. One of the sisters had heard him say to the soldiers, 'You will hurt these nuns over my dead body,' and the next morning he was found dead, hanging from a rafter in his cottage. The sisters themselves must now fear for their own lives, knowing that any hour they also may be attacked."

While the Reverend Mother paused to take a sip of water, we sat in stunned silence. Could this really be happening? To a group of our own Poor Clare sisters? On the peaceful grounds of a cloistered monastery? Could such rampant disregard of sacred property and innocent life be occurring in 1961? Just a few hundred miles off the coast of the United States of America? Shock, fear and dismay registered on all of our faces. We waited anxiously for Mother Margaret Mary to continue.

"So I have concocted a plan," Reverend Mother continued. "I have phoned the sisters in Havana summoning them to a special meeting which requires their presence in North America. This is the cover-up we are using in order to find a way to get them out of their country and here to New Orleans. The sisters are now making secret plans to leave, but they do not know yet how they will be able to accomplish this. The soldiers may not allow them to leave the grounds for any purpose, including a church meeting, so we don't know yet if the ruse is going to work or even if the sisters will be safe long enough to execute the plans we have made together. All we can do is pray for their safety and for their speedy arrival here at our monastery."

We left the meeting chastened. Some of us went immediately to the chapel to pray for our sisters. But even as I knelt asking God to allow the Cuban Poor Clares safe arrival at our monastery, I felt serious misgivings. "Now my peace is going to be disturbed," I thought selfishly. "When these sisters come in, out will go my times of special union with God. Their being here will destroy the quiet and tranquility."

Over the next few days, Reverend Mother kept giving us the latest word—"The sisters' friend in Havana just called again . . . she has been able to find a cattle ferryboat owner who has agreed to take the nuns from Cuba to Miami . . . it will be a dangerous trip and most uncomfortable . . . each sister will be able to bring only three small duffel bags with her . . . four nuns are going to stay in Cuba, hoping to save the beautiful three-hundred-year-old monastery and the sacred objects and relics, hoping that someday all the nuns can go home again." Everyone was so excited. Some of the younger nuns began running around with Spanish books trying to learn enough conversation to be hospitable.

I, on the other hand, continued to think this was not such a good idea, even though part of me felt sorry for the sisters who were losing everything. And when Mother told us younger nuns that we would be giving up our bedrooms to the new arrivals, I was even more disturbed. "Many will be sleeping in the hallways and in other locations," she informed us. I learned that my bed would be on the floor under the table in the art room! "There goes my quiet peace and union with Our Lord in my nice little room," I pouted to myself. I was fast becoming most aggravated.

A guilty conscience soon started to plague me, though, and I, albeit quite feebly, began to show interest in the proceedings taking place to extricate the Poor Clare community from their peril. Cornering Sister Anthony, a novice from Puerto Rico, I asked her to write on a piece of paper the Spanish equivalent to "Welcome to your new home." I did want to be able to greet the sisters when they came, to show them my love in some way. For days I took the little piece of paper with me everywhere I went. "Bienvenida a su neuva casa. Bienvenida a su neuva casa." I practiced again and again to be ready for the sisters.

Such excitement there was the day the Cuban sisters finally reached New Orleans. All thirty-six of us in the monastery lined up to meet our thirty new sisters from Cuba. I immediately called out first to one and then the other, "Bienvenida a su neuva casa," to which each, thinking, of course, that I really did know their language, responded by beginning to talk to me rapidly in Spanish. All I could do every time was repeat, "Bienvenida a su neuva casa!"

I found myself over the next few days coming to like many of the short little Cuban sisters; they were so happy to be alive and in a safe place and so concerned for the four nuns who were still trying to hold things together in Havana. In spite of the inconveniences the sisters' presence were causing me, I couldn't help but respond to them with care and affection. Except for one. There was one visitor to whom I immediately took an intense dislike. In fact, it was not just dislike I felt. I actually detested this sister!

And it was she I found myself looking at every time we were in chapel. Crowded now, with sixty-six nuns in a space designed for no more than thirty-five or forty, we had to take turns sitting in the main chapel. The choir chapel behind the altar, separated from the main chapel by a wall, was used for the overflow. (If it were your turn to sit in the choir chapel, you could not see the mass, of course, because of the wall; you could only hear it.) When I did get to sit in the main chapel, my seat was much farther back in the chapel. And seated directly across from me on the opposite side of the aisle every time I attended mass in the main chapel was this detested sister. My antipathy for this person shocked me. To dislike someone so much was for me a very unusual experience. It seemed almost diabolical. So not only did I not like her but I did not like how I felt. This was no way for a child of God to feel toward another. "Why have you singled out this one person to be the focus of such hateful feelings?" I chided myself. "What you are feeling for this stranger is evil." Yet struggle as I might, I could not stop the hateful feelings.

Then came the day Mother Margaret Mary gathered us all together—the New Orleans community and the Cuban sisters. I was standing midway up the stairs that led from the chapel on the first floor, waiting to hear from Reverend Mother. Looking down, I could see everybody standing there, all the Cuban and all the American sisters, everybody all mingled. All at once my eyes fell upon that distracting, repulsive nun—the one I saw all the time in chapel, the one I didn't like to look at.

"I've brought you together today," began Mother Margaret Mary because I want to put some of the Cuban sisters with the American sisters to help out with the various duties around the monastery. At that moment, the Reverend Mother looked at me. "Oh, dear God," I thought, "what is coming?" "Now the art room. Who do you want to work with you, Sister Bernadette?" With my stomach churning but with my head turning, I found my eyes singling out the very nun whom I had been trying so hard to avoid. My whole being was drawn toward that one whose presence had created in me such a puzzling negative reaction. "Conquer that bad feeling you must," I said to myself. "Choose this sister that for some strange reason you find so repulsive." "Sister Rosa," I called out to the Reverend Mother. "I'd like Sister Rosa to help me in the art room."

At that moment—that very moment—something extraordinary happened, something I have never been able to explain. My hatred suddenly turned to love. Total love. Unconditional love. Love which during the hard years that lay ahead was to help me make the passage.

Only decades later was I to come to suspect that evil forces had endeavored to prevent my association with this beautiful soul, Sister Rosa, the nun who was eventually to be such a help to the Cuban foundation, such an encouragement to me, a supportive companion, a devoted religious, and a great incentive for us to accomplish so many unexpectedly wonderful things for the Cuban community. Outside the members of my own family—my mother, father, sister, and brother—there was to be no human being for whom I have had more love than Sister Rosa. Ironically, there is no human being who has caused me more pain. How one person could have had such a tremendous influence is difficult to understand unless it is seen as part of God's Plan.

Sister Rosa was more than willing to do everything she could to help me in the art room and library or with any of my other duties and responsibilities. And the more I was with Sister Rosa the more I loved her. Often as I pondered this surprising development, I would hear the words Mother Margaret Mary had spoken so many, many years ago—could it really be ten years by now?—when I stepped across the threshold in the middle of the night to enter the New Orleans cloister:

"My child," she had said, "you are crossing the threshold into God's own house for love. For love of your crucified Lord and Savior. If you always keep that one intention in all your thoughts and actions you will receive the reward of that love, when you take your first step into that heavenly home and behold the beatific vision of your Spouse and Savior."

Now it seemed to me that those prophetic words had already been realized because I knew I had stepped into God's own house and it had been for love and for love of Him. And I knew that when Reverend Mother Margaret Mary looked up as I was standing on the stairway surrounded by the Cuban community—when she looked up and said, "Who do you want?" and I answered, "I want her," the one that was causing me such inner turmoil—I knew it was at that moment that I had had the grace to keep that one intention: love for God and God's own house. And this love that I was feeling now for Sister Rosa was an earthly reward for keeping that one intention. I wasn't even having to wait to enter heaven!

Because of Sister Rosa's willingness to help me in any possible way I began to feel an obligation to return her kindnesses. But Sister Rosa never wanted anything for herself; she always pointed to some need for her Cuban sisters. "If the sisters could have prayer books written in Spanish . . . would it be possible to make the coffee a little stronger . . . could we please find a map of California?" *California* and *fruit trees* were words the Cuban sisters used almost daily. Wouldn't this be a good place for them to establish their new foundation? Hadn't someone told them California with all its fruit trees looked like their beloved Cuba?

And at every opportunity, Sister Rosa asked me to go with her down to the basement where the Cuban sisters' small canvas bags were stored. She needed an inventory. Three bags per sister, ninety bags total. And not a single one of them was filled with personal belongings. The sisters in their escape had brought only the clothes on their backs in regard to themselves personally. What had they brought instead? Ninety bags full of things to help them set up their new foundation: altar cloths, church vestments, sacred relics, even dishes and other household things to sustain themselves until they could get reestablished. No one had known when she left Cuba to what kind of situation she would be coming. "Para la fundacion," Sister Rosa said again and again. Every time we opened another of the canvas bags— malletas, Sister Rosa called them—to inspect the contents, she would repeat as we pulled out item after item, "Esto es para la fundacion." I heard that so much I could say it in my sleep. Then there would be a flurry of Spanish with the by now famous words—*California* and *fruit trees*—sprinkled in followed by a plaintive, "You help? You help with the fundacion?" "Oh, yes, I help," I'd say just to pacify her. Then I got to where I would say myself "Para la fundacion" when we pulled an especially beautiful altar cloth or a carved statue from one of the malletas. Only later did I realize how seriously Sister Rosa took my responses. She assumed I really was going to be part of this foundation that they were going to make in America. But for me this was the furthest thing from my mind; actually I was humoring her.

As the months went by I began to worry that in our love and support for each other Sister Rosa and I were breaking an important spiritual code. No religious was to have "particular friendships"—all the spiritual books explicitly taught against this. You were not to be seen with the same person all the time; you were not to be partial one over the other; you were, instead, to spread your interest and participation around to everybody. Evidently, Sister Rosa was plagued with thoughts about the same thing, for one day she brought a book to show me— Tanquerey's *The Spiritual Life* written in Spanish—the same day that I had brought a book to show her—Tanquerey's *The Spiritual Life* written in English.

We sat down in the library and read together:

> Friendship can become a means of sanctification or a serious obstacle to perfection according as it is supernatural or merely natural and sentimental in character. We shall treat then: (1) of TRUE friendship, (2) of FALSE friendship, (3) of that friendship wherein there is an ADMIXTURE OF THE SUPERNATURAL AND THE SENTIMENTAL.

From this reading we came to an intelligent and deep understanding of our relationship with one another and our relationship with God. We were to keep God first; we were not to come to depend on earthly love and assistance; we were not to exclude anyone; our commitment was to realizing God's desires and intentions.

"Let's make a pact," I urged Sister Rosa. "Every evening after night prayers we will meet back in the art room, get down on our knees, and say together three Hail Marys. This will show our intention to keep perfect purity in our love and relationship with one another." So began the practice which was to last long after Sister Rosa—and later I—left New Orleans.

CHAPTER ELEVEN

A MIRACLE OF GRACE: GOD HAS HIS WAY

> Commit your life to the Lord,
> trust in him and he will act,
> so that your justice breaks forth like the light,
> your cause like the noon-day sun.
> FROM PSALM 37

One day we received word that the remaining four sisters who had stayed on at the monastery in Havana would soon be leaving. The fighting in the streets was continuing; the soldiers were still bivouacked inside the cloister; the sisters continued to fear for their safety. "Let's hope they get out soon," we American sisters would say to one another, "or there may be no reason to keep on expecting them."

When the sisters did finally arrive, their stories mesmerized us. . . .

We were in the chapel saying the Divine Office . . . there were soldiers standing in the doorways with their guns; others were milling all around the monastery . . . suddenly bombing began, very close by; the whole building shook with the loud sound . . . we kept on calmly praying, paying no attention to the noise whatsoever . . . afterwards one of the soldiers, visibly shaken, remarked to one of us, "I don't know

*how you could do that! Weren't you afraid?" After that many
of the soldiers would surreptitiously come up to us and ask
for a rosary or a medal.*

One story would recall another. Now that the sisters were
reunited, they told their stories with relief and obvious grati-
tude.

*Remember the night the revolutionaries woke us up. "Get
immediately down to the chapel, all of you." "But Sister
Natividad is an invalid; she cannot come." "Take her down
with you anyway; all of you must go to the chapel." That was
the night the soldiers searched for every accounting record
they could possibly find, every book housed anywhere in the
monastery. They even took every pen found in our rooms or
anywhere else in the building. Yes, and then they built the
big bonfire out in the middle of the courtyard and burned
everything they had gathered.*

Then at this point one of the Cuban sisters would start
laughing.

*Remember when the soldiers were looking for cash . . .
remember how Sister Rosa tricked them by wiggling and
giggling?*

It seems that the soldiers had ordered all the sisters to stand
in a lineup where they would be frisked to see if they were hiding
any money. Well, naturally, some of them were, in anticipation
of the Order's trying to get out of the country, particularly Sister
Rosa who had been named informal treasurer.

Sister Rosa had made a money bag to go around her waist
under her habit, and not only did this money bag make her bulge
out in funny places but it also greatly hindered her walking.
"Wobbled was more like it," Sister Rosa would laughingly say
when someone was telling the story. "I had to walk very carefully."

On this particular occasion with one soldier holding a gun on
each sister, another would frisk her hoping to find money. When
the search reached Sister Rosa and the soldier started to feel her
all over, she went into an apoplexy of giggling and wiggling.
"Ticklish . . . oh, I cannot stand it, I am so ticklish," Sister Rosa
kept exclaiming to the soldiers. "I cannot stand it." Finally, the
soldiers gave up. This silly, hysterical nun could not possibly be
responsible enough to be carrying any money. Other nuns found

equally ingenious places to hide funds they were saving for the escape . . . under flower pots in the courtyard, in shoes. It was a time for great creativity.

One of the things I always noticed when the Cuban sisters gathered to tell their stories was the kindness with which they always spoke of their captors. Even Castro, who was responsible for their exile, was not spoken of harshly. The sisters prayed for him every day, praying for his conversion. This was a constant ongoing thing with them; they did not condemn him but just prayed for him. And the sisters' love for their homeland was deeply touching. They spoke of Cuba often with great longing and affection. These storytelling times would bring the American sisters even closer to the Cuban community. We found these lovely women extremely polite, courteous, very, very helpful to us, and very grateful. It was not hard to love them.

Mother Margaret Mary put me in charge of teaching the Cuban sisters English. This brought us even closer together. I began by labeling everything around the art room with its English name. Practice, practice, practice—that is what the sisters and I did hour on end. Finally, they learned enough words that I thought we could do something like an old-fashioned spelling bee. I would give the Spanish word and the sister would be expected to respond with the English. There was much fun and a lot of giggling as we attempted to "spell down" to the winner. The sisters would also take short selections in English from the lives of a saint or a poem and read these during recreation. About the most we ever accomplished, however, were simple words and phrases, and to this day the Cuban sisters who are still alive speak very little English.

Perhaps it was the teaching or perhaps the close relationship I had with Sister Rosa, but for some reason the Cuban sisters put me up on a pedestal. They looked up to me. I was "big stuff" in their eyes, some kind of hero. This adulation was very annoying to me at first, but as time went on I came to see the good in the situation. For now, because they saw me on a pedestal, I tried to be the person they saw in me. So whether I wanted to do the right thing or whether I wanted to be full of patience or whether I wanted to do this or that—all were out of the question. I had to be what I was expected to be, and this predicament greatly helped me. When I would look within and see my weakness, all I could do was turn to God and say, "Look, God, You know what I am; You know my weaknesses. You know I can't be this thing the

Cuban sisters expect me to be unless You Yourself do it in me."
This attitude really helped me to begin to believe in myself more,
and it gave me an opportunity to become abandoned to God's will
in everything that happened.

One day I was in the office talking to Mother Margaret Mary
when the mail came and she opened a letter from Bishop
Rummel which included a check for $10,000! "To help sustain
the Cuban community," the Bishop had written. For this money
the Reverend Mother was very thankful.

Immediately the money was put to use building sleeping quarters in what had formerly been the attic. Twenty divisions were
made from the floor almost all the way up to the ceiling—one
division at each small window—and each division became a little
room that held a bed, a washstand, and a chair. A large bathroom was added with sinks, showers, and commodes. The addition was lovely when it was finished, and the younger Cuban
sisters and several Americans assigned to this area moved to our
new rooms with gusto.

In spite of the greatly improved living conditions, the Cuban
sisters did not stop talking about their foundation. In fact, after
the arrival of the last four sisters to come from Havana, there
were many more trips down to the basement, many more in-
spections of the malletas, many more authoritative observations:
para la fundacion. The Cuban Mother Superior, Madre Clara,
spoke often to her charges about California and the fruit trees—
conversations Sister Rosa would pass on to me—and then suddenly the name Bishop Garriga began to crop up frequently.
Evidently Mother Margaret Mary, in deference to the insistence
of the Cuban sisters that they be allowed to continue their own
foundation, had contacted Bishop Garriga in Corpus Christi.
There was already a small group of Spanish-speaking nuns
living in Corpus Christi, so perhaps this would be a good place
for the Cuban Order to go.

Now I began to take seriously all the talk about "para la
fundacion," my previous egging along of Sister Rosa, and my
humoring her. "Sure, I'll help you set up your foundation in a
new place" took on new meaning. "What am I promising?" I
would now ask myself. "What am I saying? I'm happy here in
New Orleans; I have my grave site all picked out; I don't want to
leave here." However, this constant review of the contents of
those malletas to see what in them would be a help for the new
foundation and then all the careful packing and repacking began

to give me a feeling, unconscious at first I think, that I belonged to all of this, that I was a part of what was to happen.

I certainly enjoyed singing their Spanish hymns; I enjoyed sitting in the back of the chapel or in the parlor when they had a Spanish conference or a Spanish visitor. I enjoyed listening and being a part of all of that. So I found myself ambivalent as the preparations became more serious for Mother Margaret Mary and Madre Clara to go to Corpus Christi to check out the situation. I didn't want to leave New Orleans, yet I felt I was supposed to be part of what was going to happen in Corpus Christi.

But then came the bombshell. I was not chosen.

When Mother Margaret Mary called everyone together to announce who would be going with her and Madre Clara to check out the situation and if feasible to help the Cuban community get started in Corpus Christi, my name was not listed. I shall never forget that moment. I can still see myself sitting in the library with everyone else—the library was our community gathering room for such occasions—sewing on something. Mother Margaret Mary began to tell us about the prospects for the Cuban community at Corpus Christi. "Madre Clara and Sister Rosa will be going to represent the Cuban sisters and I together with"— and here I held my breath because I knew my name would be the next one—"Sister Mary Francis Clare and Sister Dominic will join them." This was one of the rudest awakenings of my life! Of course, upon hearing the names of those who would be going— mine not among them—everyone in the room turned and looked at me. I remember trying to put on a very nonchalant appearance, but the truth was I was shocked. I really felt stunned because somehow in my heart I knew that I belonged with the Cuban community. I had this gut feeling that I was to be one of them, so this turn of events was very puzzling. The Reverend Mother's decision was just something I was not ready for, something I didn't want to hear.

"Stop this sniveling," I charged myself over the next few hours and days. "You know that Mother Margaret Mary speaks in God's place. Your superior's will is God's will for you. About that you have always been certain. So there is God's will in what the Reverend Mother is saying." So I bore up with it as best I could and continued to help Sister Rosa and the Cuban sisters make the necessary little preparations so they could get ready "para la fundacion."

Then on August 15, 1962, the day Sister Rosa was going to get into the car to leave for Corpus Christi, driving out of my life forever, she told me something astonishing. Or I think, based on the little Spanish I by then could understand, that it was astonishing.

"Walk with me over to the grotto," Sister Rosa said to me a few minutes before the scheduled leaving time. (Shortly after coming to the monastery ten years ago I had urged Mother Margaret Mary to let the gardener and me build this rough cement grotto which housed a statue of the Blessed Virgin Mary.) The Cuban nun's face was calm and peaceful. "How can she look so happy and even joyful," I asked myself, "when we are about to be separated forever?" But as we walked, Sister Rosa's words began to throw some light on the situation.

"I have had a vision," Sister Rosa told me. "I have seen the Blessed Mother. You will be closely linked to the Cuban community." That was all I could make out, and I never asked Sister Rosa for more elucidation. Although years later, after Sister Rosa died, one of the priests in Corpus Christi said to me, "Oh, I wish I had known that sister—the one who had the vision of the Blessed Mother." What I did know was that Sister Rosa was now completely at ease with my not going at this time to Corpus Christi. "I have had a vision," she said repeatedly. "I have seen the Blessed Mother. You will come to Corpus Christi."

For whatever comfort Sister Rosa's vision had given her, I had no counterpart in my own experience. Sister Rosa got into the car with Madre Clara, Mother Margaret Mary, Sister Francis Clare and Sister Dominic. All of us stood there waving to them as the car backed out of the big gate onto Magazine Street. I had such a strong inner feeling that I was a part of this Cuban foundation's getting a new start in America; yet I had not been selected. "All I can do," I said to myself, "is throw myself into extra vigilance and prayer and resignation in accepting whatever comes." I knew by now that it was one thing to say, "Yes, God, I will do Your will," but it was another thing to try to bring yourself really to mean in your heart what you are saying with your lips. I knew this was what I had now to accomplish.

That night after prayers, Sister Holy Spirit, the niece of Madre Clara, came up to me ever so timidly. "Come go to art room," she spoke haltingly. "Come . . . kneel . . . say three Hail

Marys." Tears gushed down my cheeks as we knelt there praying. I kept asking myself, "How could this quiet little Cuban nun have known so clearly what I needed?" From that time forward I felt very close to Sister Holy Spirit.

In a few days Mother Margaret Mary returned to New Orleans, but she reported to us not a word about Sister Mary Francis Clare, Sister Dominic, and the Cuban sisters. Those of us who were interested in knowing how they were doing had to wonder in silence.

My thoughts were constantly on the sisters. Everywhere I looked someone important to me was missing. In the art room, at night after prayers, of course, it was Sister Rosa. But even at work during the day I was constantly confronted with the absence of those gone to Corpus Christi, for it had been Sister Mary Francis Clare who helped me publish the *Unity* publication which went to the Poor Clares in the federation.

Then one day, lo and behold, out of the clear blue sky who should appear on the scene but Sister Mary Francis Clare. Not a word was given in explanation for her return nor did we hear anything about how the Cuban sisters were doing. We also did not know what happened to Sister Dominic; much later we were told that she had become ill and would not be returning to the monastery.

"How in the world are the Cuban sisters making it in Corpus Christi?" I kept asking myself. "They have never lived outside the enclosure, they don't know the culture, why, they don't even know the English language!" My heart was crushed thinking about those precious nuns from a foreign country—the two who went with Reverend Mother originally and the eleven who had followed—living now in a strange city all by themselves, trying to get their foundation reestablished.

Finally, my curiosity and concern got the better of my discretion. Breaking the customary silence we sisters kept when working close together, I blurted out one day while we were kneeling out in the garden weeding, "Sister Mary Francis Clare, who is helping the sisters in Corpus Christi right now?" She shot back her answer, crisp and definitive. "Nobody, and there are no intentions of sending anyone there to help them."

"This is your cue," I tutored myself the rest of the day. "If you feel so bad about all this, make an effort to do something about it.

Go to Mother Margaret Mary and ask her if *you* can go to Corpus Christi to help the Cuban sisters."

———————

"Absolutely not." That was Mother's answer. There were no questions or discussion about it.

"Get it out of your mind," I cautioned myself as the desire to go to Corpus Christi continued to plague me. "Mother has said no, and that is that."

"But you could go ask again," another insistent voice spoke inside me. "Isn't it possible that God sometimes uses *desire* as the impetus to point us toward where we ought to be going?"

Finally, several weeks later I succumbed. I went back to Mother Margaret Mary's office. "Reverend Mother, I need to talk to you," I began very tentatively. "Certainly, child, go right ahead." It was clear she had no idea I was still intent on asking permission to go to Corpus Christi.

"I have tried to put the Cuban sisters and their efforts to establish their foundation in Corpus Christi out of my mind, but the urge to ask you again to let me go help continues in my thoughts night and day. It's almost unbearable." I was now having to screw up all the courage I could muster, for clearly the Reverend Mother was not being receptive. So speaking as rapidly as I could and probably much too loudly, I asked, "Mother, may I go help the sisters in Corpus Christi?"

Again the answer was an unequivocal no. "There is no intention of sending anyone to help." This was the only answer Mother Margaret Mary would give me.

"In obedience," I scolded myself as I left the Reverend Mother's office, "you must put out of your mind going to help the nuns in Corpus Christi. You know that you have always believed that your superior's will was God's will, even when you were a little girl and that superior was your own mother." So day after day I disciplined myself to not think about the sisters in Corpus Christi.

———————

Then one day it happened. Perhaps the single most important spiritual event of my life, and certainly the event that changed the direction of my life forever.

We were in chapel for the mass. Since eighteen of the elderly Cuban nuns were still with us in New Orleans awaiting the call to join the sisters who had gone on to Corpus Christi, we were still very crowded. This meant that we were still rotating—some sitting in the small main chapel in front of the altar, some sitting on the side, and some sitting behind the wall in the choir chapel. Today it was my turn to be seated in the choir chapel; here, of course, because of the wall, we could only hear the mass. We could not see it.

Since I could not see the priest's movements during the mass and since many of his rituals were carried out in silence, I had only an inkling of the time that was passing. But then the little bell tingled, a signal that the Host was about to be raised. Suddenly—and to this day I still cannot explain what happened—I saw the Host being raised in the priest's hand. Just as if the wall were not there. I saw the Host, and at that same moment I heard words—very audible words. It was not as if they were spoken outside but from somewhere within me. The words were very clear, just like someone speaking. "Why are you afraid to come to Me in Corpus Christi?"

But there was something strange about the way this question was spoken. It was as if the words "in Corpus Christi" were in parentheses. "Why are you afraid to come to Me (in Corpus Christi)?" Now, how it is possible to hear parentheses in speaking, I do not know, but I knew there was something different about the words at the end of the question. The only thing I could figure out was that since the words *Corpus Christi* meant the "Body of Christ" as well as the name of a city, I was being reminded that the body of Christ and the work in Corpus Christi were related. (Years later I was to realize another explanation.)

I knew without a doubt that this voice was Christ speaking to me. And I flashed back an answer.

"Lord, that is a funny thing to ask me. You know that You have already spoken to me through Reverend Mother."

With that response, I let the matter rest.

In a few moments it was time for those of us seated behind the wall to go out into the main chapel for Holy Communion. Upon returning to my seat, just as I walked under the archway of the main chapel to go behind the wall to the choir chapel—right as I passed through the door—I hear words again. "I am your strength," the voice said to me. "I am your strength." Oh, how over the years I have pondered those words. Not "I will be your strength," but "I am your strength."

Somehow I made a connection between the words I heard during the elevation of the Host and now these words. "Okay, Lord," I answered, "since you are my strength I will ask again to go to Corpus Christi."

Immediately after chapel I went to my room, found a little card and wrote the request:

"Dear Mother Margaret Mary,

If it is God's will, may I go to help the sisters in Corpus Christi?
                              Sister Bernadette"

Slipping into the Reverend Mother's office when no one was looking, I put the card on her desk. I had no idea what would now happen.

I had put the request on her desk on Sunday. Early the next week, Mother Margaret Mary called me into her office. "Here is your airline ticket," she said, showing me a packet. "On December 12, you and Sister Veronica of the Cuban community will be flying from New Orleans to Corpus Christi." It wasn't until we were in the tiny plane, bucking the winds as we headed for Corpus Christi, that I took note of the date, December 12. It was the day for celebrating the Feast of Our Lady of Guadalupe.

The Cuban sisters arrive in New Orleans from Havana. Sister Rosa is top row, fourth from left. Caridad de Velasco is second row, sixth from left. Madre Clara is front row, third from left. The statue in the back is Saint Clare.

The Poor Clare Monastery in Havana which the Cuban sisters had to leave when their lives were in danger during the Cuban Revolution.

The Cuban sisters loved the arches of the cloister of their monastery in Havana.

The altar in the monastery in Havana. The Cuban sisters made beautiful hand-sewn pieces for use in the chapel.

The Cuban sisters play and dance during recreation period in Havana.

How the Cuban sisters en-
joyed the garden areas of
their monastery in Havana.

SECTION SEVEN

CORPUS CHRISTI, TEXAS
1963-1985

My Life With the Cuban Sisters

CHAPTER TWELVE

STRUGGLES, COMEDY AND SURVIVAL

> Your will is my heritage for ever,
> the joy of my heart.
> I set myself to carry out your will
> in fullness, for ever.
>
> FROM PSALM 119

I walked up the steps and onto the porch of the rented house on Third Street, the quarters which Sister Francis Clare had eventually found for the Cuban sisters after viewing their original temporary dilapidated quarters on Antelope Street. At that moment, I felt the same affirmation that I had experienced that night ten years ago when I stepped across the threshold of the Poor Clare monastery in New Orleans. This place was exactly where I belonged. I felt exuberant.

Madre Clara and Sister Rosa led me to the place where I was to put my bags down. I was touched. Even though the Cuban sisters slept four to a room with only a curtain around each bed for privacy, they had prepared for me a tiny private room at the top of the stairs. I knew this was one way they were letting me know how much they appreciated my having been given permission to come and help them.

Standing in that little room, I suddenly felt the duty and responsibility that came with my charge. I understood clearly that I had been given permission to stay with the Cuban sisters for only one year, after which time I would either become a member of their community or return to my own Poor Clare monastery in New Orleans. I also understood clearly that whether I would stay here or return depended upon how stable the Cuban foundation was at the end of the twelve months I had been given and upon agreement between the two monasteries.

Only one year to get roots firmly planted. And here I was, not even a voting member of this community and only able prudently to make suggestions. "Just remember what you have done in the past," I admonished myself as I unpacked my bags. "Remember what worked when the sisters were in New Orleans. Try to be helpful without being pushy about it." Standing there, thinking about the uncertain future, I realized how much I wanted to instill confidence into these sweet sisters who had so many handicaps working against them. I wanted with all my soul to help them find the courage to act.

"Make a sort of game out of it," I tutored myself as I left my little bedroom to join the others on the bottom floor. "Get them started on a task, and when they catch on, you can back out and leave them on their own." Today I thank God that at the time I did not realize what a job I would constantly have on my hands!

I later would begin to see the scope of the task that lay at hand. But at first the situation was only funny. . . .

Sister Rosa was addressing Christmas cards. "Look over . . ." she asked me, holding out the envelopes she had already completed. "Look over. See if right. . . . Want to thank for helping us."

I took the envelopes and began to check them. Sister Rosa had copied the addresses from the envelopes that the monastery bills had come in. There was a card for the light company, a card for the Corpus Christi water system, a card for the garbage service, each address painstakingly reproduced in Sister Rosa's awkward printing. "These are all fine," I told Sister Rosa, "you're doing a good job." But then I saw it—the card for the Corpus Christi Produce Company. On the envelope Sister Rosa had carefully copied:

    Corpus Christi Produce Co.
    Potatoes, Tomatoes, Onions, Apples, Oranges
    238 N. Port
    Corpus Christi, Texas

"I think," I said to Sister Rosa, with as straight a face as possible, "that we will have to do this one over."

Madre Clara came into the room and handed me a grocery list, obviously relieved to pass the task of grocery shopping on to someone who could speak the same language as those people at the StopnGo down the street. "You . . . Sister Rosa . . . go store . . . need store."

I read the list on the torn piece of paper the Madre Clara handed me:

    Foo for dog
    Oats for Father—Quick

Sparkling Pots and Pans
1 chore girl
1 brite and shiny
2 potatoes cheap

This time I couldn't keep a straight face. I just couldn't help it. I had to laugh at the sister's misunderstanding as I deciphered the Madre Clara's grocery list, trying to figure out what she meant:

Dog food
Quick-cooking oats
Joy liquid detergent
1 box Chore Girl scouring pads
1 can silver polish
2 bags of potato chips

Even now, twenty years later, the sisters still copy words from the label of something they want from the store and at least half of the time what they copy is not the name of the item but some advertising slogan or product description. Over the years we have learned to joke about this, and it is a constant source which stirs my love for this tiny band of courageous women, women who in the middle or latter parts of their lives had been forced to flee their beloved home with little more than the clothes on their backs, women who can now have only in memory the grassy lawns, sweet-smelling flowers, and blossoms of orange trees that surrounded their spacious monastery in Havana. Women who lost forever the opportunity to continue their dedicated prayerful lives in a cloister that for more than three hundred years had stood for all Cubans as a place of constant prayer, a place of adoration of the Blessed Sacrament and devout love for the Virgin Mary.

In Cuba, of course, as cloistered nuns, the sisters had never left the enclosure for any purpose—certainly not for shopping—and now here as refugees in America they were suddenly thrust into the very thick of daily living. I kept reminding myself of this fact when it took much time and questioning to discover that the "marbles" the Madre Clara was asking for were not really marbles but little marshmallows for the salad. Or that "live chunks" meant liver chunks for dog food.

Sometimes adding some motions helped bring understanding, but this method of communication could also bring catastrophe.

The garage apartment out back needed to be vacuumed in preparation for setting up an altar bread-making machine the sisters hoped would be a way for them to make a living. "You know, use the vacuum cleaner—zizzzzz" was the best instruction that could be given in English. To this instruction, however, was added a motion back and forth and all around, to illustrate vacuuming.

"O, si, si!" replied Madre Clara with a knowing smile as she headed off to do the job. About a half hour later, streams of water were gushing down the stairs of the garage apartment and flowing all over the ground. Madre Clara had misunderstood the action. Instead of using the vacuum cleaner, she had promptly turned the water hose on full blast and flooded the entire apartment.

But then there were times that Sister Rosa knew exactly what to do. "Come with me," she said to me one morning. "We go get Coke." After she picked up two six packs of empty Coke bottles, we headed on foot to the corner StopnGo.

There I watched a very confident Sister Rosa walk proudly into the store and announce to the clerk, "We get more Coke. We bring empty bottles for you." She then put down the empties, picked up two full six packs, and proudly marched right to the front door, leaving me dumbfounded and the clerk scratching his head in wonderment. Explaining the problem to the clerk as best I could, I hurriedly chased after Sister Rosa. "We have to give the man money," I told her. "We must go back to the monastery for money." "But," she insisted on arguing back at me, "bottle say 'return empty for more.'"

Yet in spite of the difficulties, the Cuban nuns were determined to reestablish their foundation in this strange America. They were determined to continue as the ancient established community they had been for more than three hundred years. They were determined to find a way to return to cloistered lives devoted to prayer for themselves and the whole world.

And even as I struggled to know how best to help the sisters, I was deeply moved by their never-wavering intentions to keep their monastery intact and active no matter what the external circumstances. As the days and weeks went on, I felt my own commitment grow even stronger. These Cuban sisters *must* be able to survive as a community here in America.

"Why are you afraid to come to Me (in Corpus Christi)?" The words I had heard that morning at mass in the New Orleans

monastery now came back to me. "Well, I am here now," I said silently. "I have come to do Your will, whatever that may be!"

But looking around at the circumstances in the temporary quarters on Third Street in Corpus Christi and realizing that I did not know how to help this group of refugee sisters return to being a self-sufficient Order caused my heart to "sink to my boots." How could these women who spoke no English, who had no skills or experience in negotiating life in the outside world, who had no reserves and no resources except the baking of altar breads for the Diocese of Corpus Christi possibly find a way to provide even for their daily sustenance after the outside funding had ended—much less find a way to raise the money and build a new monastery on land they did not yet have and be able to carry on indefinitely? And in my position right now, there was so little I could rightfully take upon myself to do since I was not yet even a member of this community.

There was also the sad—and serious—matter of Sister Caridad . . . Sister Rosa had tearfully confided the story to me one day not long after I joined the group in Corpus Christi.

Oh, it was so long ago. Children were playing together on beach in China, parents nearby. Educated, well-to-do parents—professors, doctors. Mean men run onto beach and steal children right from under parents' noses. Take to big boat. Boat goes to Cuba where Chinese children are sold as slaves.

When grown, two of these slave-children get married and have a daughter, Caridad de Velasco, who carries the last name of the slave owner of her father. When Caridad was seventeen she came to the Poor Clare monastery in Havana. "I want to be a nun," she told the sisters. "God has called me to be a Poor Clare nun."

But she was not accepted. "You can only be a lay sister," they told her. Lay sisters do the chores but cannot wear the habit or sing the divine office. "I want to be a choir sister. I want to devote everything to the Holy Mother," Caridad responded. But the role of choir sister she was not allowed. Was it because she was daughter of Chinese slaves? Was it because her family had no dowry to present to the monastery? Was it because . . . who can say fully why she was not accepted.

But Caridad de Velasco would not leave the monastery. Nor would she take on the identity of a lay sister. God had called her to be a Poor Clare nun, she insisted, and only this would fulfill His holy will. "I will clean and cook and serve," the young girl said to the sisters, "but I will do so as one waiting to be accepted as a novice."

And for sixty-one years, she had lived in the basement of the Havana monastery, cleaning, cooking, scrubbing, serving— waiting to be accepted as a novice. Docile. Obedient in all things. Always present, if inconspicuously, for prayers.

What to do with Caridad de Velasco when it was time to evacuate and take flight to America? Nobody knew. Then Mother Margaret Mary in New Orleans said, "Dress her in a nun's habit and bring her to safety also."

So, now, Sister Caridad was here in America, dressed as a nun—with nobody except the Cuban sisters and Mother Margaret Mary knowing the true story. But she is not officially a nun. What would the Bishop do if he discovered what we had done? No woman can ever become a nun without serving as a postulant and a novice. But Caridad cannot go back to Cuba . . . she would die if she were separated from this community . . . and, any way, where could she go if she were sent back to Cuba? The monastery has been her only home for more than sixty years. . . .

I had been deeply touched by the story. I didn't know a solution to the grave situation. I didn't even know how to respond. But I knew this: Caridad de Velasco was just one more reason that I needed to do everything I possibly could to help the Cuban community. Surely God would show us the way.

It was then that I remembered so clearly that second message which I heard that day in New Orleans. ***"I am your strength."*** "Yes, Lord," I prayed silently, "You are the only hope we have." It was clear to me now as I surveyed what looked like a bleak, maybe even impossible future, that *"I am your strength"* were the only words any of us had to hold onto. With those thoughts I confidently turned to the task that lay at hand.

———————

I began to take a mental inventory of the sisters' skills. Was there anything else they could do to help support themselves in addition to making the altar bread, an activity which I knew in truth would never bring them in very much profit? I racked my brain for possibilities.

Ah, perhaps they could do ceramics. I had learned how to clean and glaze and fire by practicing on my own in New Orleans, so perhaps I could teach them. But what could we make? What items would be simple enough that beginners could make them and yet be interesting enough that the public would want to buy them?

One morning the idea came to me. For some unexplainable reason I woke up thinking about the platform Papa built each Christmas when I was home. What a happy memory. The French doors were locked for two weeks before Christmas so that none of us children could pass through. A sheet hung over the glass panes in the door to thwart any peeking. Then came Christmas morning. We were finally allowed to see. The waist-high platform on which was set up a complete little village with a train track that went all through the village and back under a great big mountain that had been built up against the side of the wall. The train would go under this mountain and come out on the other side. There was a little place in the village train station from which you could manipulate the train. You could make it go faster or slower and you could stop it anywhere you liked. All the tiny houses in the village lit up, and there were real-looking little street lights that also shined magically along the streets.

But the best was yet to come. Papa designed and built a complete nativity scene and placed it back into the big window directly behind the platform. As you stood looking at the sweet little houses and at the train running through, you could see off in the distance—as if it were just out on the edge of the village— the stable with the Holy Christ Child, Mary, Joseph, the shepherds, and even in the farther off distance the wise men on their way to worship their new-found King. Lying on my cot in Corpus Christi, I could remember the beauty of that moment as clearly as if I were ten years old again. And at the same time the beautiful scene seemed only a dream, so far away.

Suddenly, I realized that in that memory lay the answer to what we could do with ceramics. We would make nativity scenes. And after we learned how to make the pieces, we would learn how to sell them.

But where could we begin?

Someone told us about Rita Steinmetz who had a small ceramic shop in her garage. "Would you help us?" "Of course," she said, "I'll be glad to do anything I can." So off I went to get what we would need first, a large kiln and our supplies. How could we pay? "Oh, you can pay me as you go along," the generous proprietor told me. I also noticed when she gave me the bill that she had given us an enormous discount.

So the Cuban sisters and I were in business. With the help of Sister Holy Spirit, who was especially enthusiastic about the new project, I took the next step: setting up a ceramic studio in the basement of the rented house. Then I called the Bishop. Would he come over and perform a special service? So down into the basement we all trooped with the Bishop who was there to pray a particular blessing: That our kiln and supplies would be used for God's purposes and that the Cuban sisters' first little entrepreneurial efforts would prosper.

We soon became conspicuous to the town. Because we continued to make altar bread as we tried to get the ceramic business going, frequent trips to the post office were necessary to mail the bread to different churches in the diocese. I could not understand why Madre Clara always insisted that we go all the way down to the main post office, which was twelve blocks away, instead of going to the post office which was much closer to our house. Finally, I got the truth out of Madre Clara. "We no go that post office . . . man no smile . . . we afraid of him." "Well, we can put a stop to that right now," I told Madre Clara. "I'm not afraid of a man who doesn't smile, so I'll be the one who stands in the front at the counter."

We had to carry these huge, cumbersome black bags with zippers across the top to the post office ourselves, and, of course, we could do this only by walking. So there we would be, out on the sidewalk, two or three short Poor Clare nuns, heaving and dragging and pushing these bags as we went on our way. Gradually, locals who would see us carrying these big bags of altar bread began to stop and give us a lift.

This gave me an idea that perhaps we could be like the original mendicant Franciscans. We could do a lot of begging, asking for what we needed from a community that was beginning to realize that we were present. So I came up with the bright idea that we would make flyers introducing ourselves,

saying we were a mendicant Order and would appreciate whatever could be given to us to help pay our grocery bills because we
had no salary. We took several hundred of these folded sheets of
paper down to Spohn Hospital and put one in every car in the
parking lot. If the windows were open, we threw the flyer in the
window. If the windows weren't open, we stuck the flyer under
the windshield wiper.

And what a tiring, hot job. But we were finally finished and
were so proud of ourselves as we walked back down the street,
reached the house, and went in to get something cold to drink.
Almost before we could finish our tea, the telephone rang. Madre
Clara answered and engaged in quite a short conversation.
"*That* was Bishop Garriga," Mother said directly to me when she
came back into the room. "He said he had been in the hospital
and that when he came back out to his car, there he found that
flyer announcing that we were begging. He really scolded us,"
Madre Clara relayed. "We are never to do that again." But
evidently the Bishop hadn't gone so far as to go around to all the
cars and retrieve the flyers, for in a few days we began to see
results. Many people began giving.

That dubious success gave me other ideas. Perhaps we could
go to commercial suppliers directly. So down to the Borden Milk
Company I went to talk to the president and ask for free milk.
(Ben Matelski, the Borden official who gave us our three gallons
of milk each week and the extra cottage cheese for an entire year,
had a way of saying *beautiful* that stretched the word out to be
"beauuuutiful." Madre Clara, who could not remember Mr.
Matelski's name, insisted on calling him "Mr. Beauuuutiful" and
to this day I can say to her, "Do you remember beauuuutiful?"
and she will answer, "Oh, si. Borden milk man.")

Then on to the thrift store. Would they give us day-old bread?
And for the next twenty years we received our bread free, due to
the generosity of the Buttercrust Bread Company. The grocery
stores agreed to let us pick through the vegetables when they
cleared out to make room for fresh produce. And someone introduced us to Elena Kenedy of the Kenedy Ranch, a sweet, cultured little woman who became a great friend and benefactor.
Many times Sister Rosa would contact her and let her know we
needed some money and it would always be forthcoming.

And eventually someone in the community even gave us a car.
"You drive," Madre Clara uttered. "But I have no license," I
pleaded. "Get one!" was her curt reply. "If you want me to drive,"

I told Madre Clara, thinking of that disastrous day in Ft. Myers when Sister Loretta and I drove right across that lawn and directly into a huge palm tree, "please let me go take driving lessons." So off we went to driving school, Madre Clara always insisting on being along because no nun should be alone in the car with a driving instructor.

Naturally, Madre Clara insisted on being in the car the day I went for my driver's license. "No one is allowed in the car except the applicant when we give this test," the stern state trooper said as he glowered at me. Well, I knew you'd never be able to convince Madre Clara of this.

"Look, sir," I replied as politely as I could, "there is just no way she is going to let me go alone with you. We'll have to take her." "Then tell her to sit in the back seat and to be quiet," the trooper said, none too happily. So Madre Clara got into the back seat where she promptly reclined against the seat, chewing the gum of which she had become so fond since she came to America. She was sitting there totally contented.

"I want you to go about ten miles an hour, and then when I say 'stop' I want you to put your foot on the brake as quickly as you can," the trooper instructed me. "Sir," I responded, "please let me explain this to the sister back there so that she will know what we are doing." So I told Madre that I was going to slow down and that when the trooper said 'stop,' I was really going to put the brakes on. "Brace yourself," I warned her. Madre Clara replied, "Okay," and went on chomping her gum.

I began to drive ever so slowly and suddenly the man yelled, "Stop!" I guess my reflexes were better than he anticipated for I jerked us to a halt so quickly that even he lost his balance. But to make things even worse, Madre Clara landed right on top of him, just came kerplunk, right over the top of the back seat. She told me later that she even swallowed her gum!

That episode left the gentleman a little provoked, I think, for things from that time forward did not go well.

"Go three blocks down and make a right," I was instructed. As I approached the intersection, a big moving van rolled up, blocking my view. "Looks like you planted that one," I said, joking the way I always do when I get nervous. Then when we had gone no more than a couple of blocks, a little dog ran out in front of us and I had to brake for him. "Was he another plant?" I teased the officer. After a little bit of driving, I was told to park between two poles. That looked easy, and I backed in and just fit perfectly. "Okay," the trooper said, and I stopped the parking maneuver and started the drive back to the station.

We pulled back up to the building and there stood my driving teacher, just beaming. The cop got out of the car, walked over to my instructor and said, much to my shock and consternation, "She'll have to take this test again. She failed." "Failed?" I asked incredulously. "Yes," he replied, "you did not turn your head and you were not alert when you passed street corners or braked for an animal. Also, you failed to finish your park. You did not turn your wheels toward the curb after you were situated. You may take this test again any time except today." I thought to myself as the trooper walked away, "Well, mister, I may have made the errors you itemized, but I bet I really failed the test because Madre Clara fell on you!"

The next day my driving teacher, Madre Clara, and I went back and did the whole thing over. Another officer took me out, and this time I did not say one single solitary word. And when we got to a crossroads I looked as far right as my head would go and then as far left. When I parked, I turned my wheels sharply to the curb. And I told Madre Clara to get a grip that would not let go, for under no circumstances was she to come flying over the front seat, no matter what happened. Fortunately, all went well and I got my license, I hesitate to say, though, with flying colors.

From that time forward I was the official chauffeur, with Madre Clara as my constant companion. To the post office to deliver the altar bread. To the grocery store. To take the sisters to Dr. Jimenez's office or make frequent trips to the immigration office to update their papers or make arrangements when several of the sisters decided to move to a monastery in Mexico.

As the months passed by, I became more and more edgy. My one-year dispensation to live with the Cuban sisters in Corpus Christi would soon be up, and I could not tell that one single thing had been done that would really ensure stability for the sisters here in this community. "Madre," I said one day, "why don't we start looking around for land to build a monastery?" "How can we do that?" asked Madre, sounding puzzled yet at the same time very excited. "Look, Madre," I answered, "if we don't do something, I'll have to go back to New Orleans and can no longer help you. Also, even while I am here, there is little I can do since I am not legally a member of the community and have no authority to act." "I tell act!" Madre replied authoritatively. "But it is not that easy," I responded. "Okay, I tell Bishop!" And with that the Madre Clara promptly wrote to Bishop Garriga, appointing me administrator on her behalf in order to be able to take care of whatever business would be necessary for the good of the community.

That freedom to act was all I needed! Immediately off we went to look for possible land sites on which we might build our monastery. The fact that we had no money did not faze us. "I know what we can do," I told Madre Clara. "I'll find out what land the Diocese of Corpus Christi already owns by asking a real estate agent to look the information up for us. Then we'll go look at each piece of property, pick out the spot we like, and ask Bishop Garriga to give it to us."

Day after day Madre Clara and I went out. Four places this day. Five places that. Before the search was all over, we had inspected twenty-three different locations. But then the gavel fell.

Our last stop had been on a dead-end street where we had found a large lot. The Bishop's residence on Ocean Drive was in sight in the distance. While standing on the vacant lot discussing the situation, we saw the Eiermanns, who lived nearby. They invited us into their home, and we had a great visit. Bertha that day became a lasting friend and her husband, Chester, became a "member of the board" which was then and there, at that very moment, established. But while the four of us were formulating our great plans, something else was happening. No sooner had Madre Clara and I arrived back at the rented house on Third Street than the telephone rang.

It was Bishop Garriga.

"What were you doing this afternoon?" he bellowed. This was our chance. "Why we are looking for possible sites for our monastery, Your Excellency." Then came a sudden emphatic response. "I don't want any nuns living in my back yard." And that was that. We had no idea that the lot we were looking at was part of the Bishop's residence.

In the ensuing days the Bishop must have got whiff of all our other investigations of diocesan property. A few days later we had just started dinner when the telephone rang. Sister Rosa came back to the table, speaking excitedly. "The Bishop wants to talk to you, Sister Bernadette." Immediately all the forks were put down and perfect quiet reigned. I picked up the receiver in the adjoining room. This is what I heard: "Sister Mary Bernadette, why don't you mind your own business? Who do you think you are anyway—the Superior over there? You seem to know a lot about this diocese."

"I'm sorry, Your Excellency." What else could I say? The Bishop abruptly hung up. By the time I got back to the table, a big tear had rolled down my cheek and ruined the smile I had tried to force. . . .

I was suddenly back in the Allegany convent, sixteen years old, a new postulant. So far I hadn't been homesick at all, for Mama wrote at least four times a week, keeping me in touch with everything that was happening at home. Her letters really held me together. But then a very traumatic thing happened.

One of the novices became very ill and was put in a small dormitory, isolated from the rest of us, right next door to the dormitory in which I slept. We all wanted to go in and visit her, but, of course, were not allowed to. Because I wanted so much to do something that would comfort the sister, I came up with a bright idea. "I have something in my trunk in the basement which will cheer her up," I said to myself. "The first break I have today I'll go down and get it."

In our trunks in the basement we kept everything related to our families, everything private. None of this were we allowed to have with us. I had remembered a real treasure. In a folder I had a collection of pen and pencil drawings that Bruds, my little brother, had done. Bruds was extremely artistic, and these drawings were very beautiful. Of everything I had stored in my trunk, I think these drawings were to me the most precious.

All excited, I ran up the stairs as fast as my new religious decorum would permit, and I knocked on the door. Sister Anna Marie opened it and stood there with her arms folded, looking as grim as usual. I spoke, "Sister, would you please see that Nora gets these so that she can look at them. I'm sure she will enjoy them." Sister Anna Marie reached for the folder which was in my hand. "These are things my brother drew, and I know Nora would like to look at them. Will you just be sure that I get them back because I don't want to lose them."

Sister Anna Marie looked at me from her stance about three feet away and then she took the folder holding my precious drawings and threw it right through the open door so that the drawings landed all over the floor beside me.

My heart was broken. I stooped down and gathered the drawings up and, oh, I just felt as if something sacrilegious had happened. I treasured the drawings so much . . . they were things my brother had sent me. I went back down those stairs, down to the trunk room— it was dark in the trunk room—and I opened up my trunk, put the pictures in, closed up the trunk, and then I sat down on the trunk and cried and cried and cried. That was the first time I had cried since I entered the convent. My heart was broken, and I just couldn't understand. I cried and cried, calling out, "Mama, Mama." All I wanted was to be back at home. I felt so alone and so devastated.

It was a long time before I learned that the sick sister had had tuberculosis and that everyone who knew was frightened that the disease would spread to the entire convent. Nothing of course that went in to the sick girl could come back out except as garbage. But at the time I hadn't known that and the hurt had been horrible.

"What happened?" asked Madre Clara, bringing me back from my reverie. All eyes were upon me as I answered. "The Bishop bawled me out." Then I repeated the conversation.

"But I wrote and told him I was appointing you business administrator to help us out," Madre Clara asserted. "I cannot understand why he would say to you that message." Slowly, we all returned to eating dinner.

That evening when we were at the table, this time eating supper, the phone rang again.

"The Bishop wants to talk to you, Sister Bernadette."

Again the forks went down, and I got up. What I heard this time was this:

"Sister Bernadette, I want to apologize to you. When I was chastising you this morning, I hadn't opened my mail yet. All the while sitting right here on my desk was a letter from Madre Clara appointing you business administrator. I am very sorry, and I apologize."

"You don't need to apologize to me, Your Excellency," I said in reply. But this time I went back to the table with smiles and no tears. I cannot remember what we were eating for supper, but I know that it tasted to me like food from the angels.

But that was not to be my last encounter with the Bishop. It was now near the end of my one-year dispensation. It would be decided soon whether I would return to New Orleans or remain with the sisters in Corpus Christi. Sister Mary Francis Clare came over from New Orleans to go with Bishop Garriga and Madre Clara to Robstown, about a half hour from Corpus Christi. The Bishop was suggesting that an abandoned hospital in Robstown would make an excellent monastery for the Cuban sisters, though it was in poor condition. I was not permitted to accompany the Bishop, Sister Mary Francis Clare, and Madre Clara, but I learned immediately upon their return that neither of the women was impressed.

But the Bishop seemed intent that the Robstown Hospital would become the home of the Cuban sisters. It seemed that the diocese had purchased it and it had become a sort of white elephant. Knowing that my time was fast approaching to make a decision as to where I would spend the rest of my life, I wanted to see this Robstown Hospital. I persuaded Madre Clara to ask the Bishop if we could take a ride over and see the building. She and I went down to the chancery office early one morning and caught Bishop Garriga as he was getting out of his car. I don't remember how we persuaded him to let us go, but he finally said, "All right. You go. But do not get out of the car."

So Madre Clara and I drove to Robstown, stayed in the car, and gazed upon this dilapidated building that had absolutely no yard. There wasn't even a place to hang out clothes. Well, I knew right away that no contemplative group could persevere in surroundings like that. I knew that the sisters would not be able to thrive here. I knew that there would be no reason for me to stay to help them try to build a new foundation.

Sitting there in front of the dilapidated hospital, I began using every persuasive word I knew to make Madre Clara understand that this was no place for us and that we must speak up and say so. This was the moment of decision. Either Madre Clara acquiesced and accepted the Bishop's solution and I returned to New Orleans because there would be no foundation to build, or we, with God's help, would be able to stand our ground in the knowledge that this location was not the place for the Cuban sisters. We sat praying in front of the empty building for many minutes.

When we got back to the house on Third Street, Madre Clara said, "I know what is right thing to do . . . come with me to telephone." When we got into the room, Madre dialed the Bishop's number. "Thank you just the same," she said to him when he answered, "but we do not want any part of the Robstown Hospital. We must build our own monastery." That, of course, did not endear me further to the Bishop. I was just an interfering person, influencing Madre Clara when the Bishop wanted us to take the hospital so badly. But the episode did result in guidance and direction for me. I knew now that I would be staying permanently with the sisters in Corpus Christi to help them build a monastery. I wrote the appropriate authorities to ask permission.

Time continued to pass. The Bishop said nothing about another piece of land. Then news trickled down to us through one of the members of the Catholic Daughters. The Bishop had announced that he was going to give us five acres on Saratoga Boulevard. Yet more time passed and still we heard nothing.

In order to bring things to a head, I wrote the Bishop asking him to inform us of the exact boundaries of the land on Saratoga since someone was donating some young trees and we wanted to plant them. Again the gavel came down.

"You can't have that land," the Bishop replied. "There is a problem with the drainage."

This was a great disappointment. Now it was really beginning to look as if the Cuban sisters' chances for survival as a community in the United States were slim indeed.

We went ahead, however, with what we could do. Our ceramic business was growing. I had come up with the idea of building a big black shadowbox and attaching it to the wall of the front porch at the house on Third Street and displaying in this box the all-white nativity scenes the sisters were now making. "This display case reminds me," I told Sister Rosa, "of what my brother said when he saw the trunk Mama had bought for packing my things when I left home to enter the Allegany convent. "My goodness," Bruds said when he came in the house the first time and saw the open trunk in the living room, "That looks like a casket sitting there!'"

Cissy Farenthold, who later ran for governor of Texas and whose mother owned the house we were renting on Third Street, bought one of the first nativity scenes and began to make connections for us, including arranging for an exhibition of our work in a large room she rented at the Driscoll Hotel. To make this a big event, I wrote to all the Poor Clare monasteries and told them we were going to have this exhibit and would they please send us anything they had which would be representative of Poor Clare work throughout the country. Just about everyone I approached sent something—art work, Christmas cards, oil paintings. To these we added our ceramics. The show was a great success. We made one very lovely piece—a hunting dog with long hair holding a pheasant in his mouth—that someone bought as a present for Bing Crosby's son. The exhibit had helped us immensely to be known in Corpus Christi.

But the income from the ceramic business would never be enough to take care of the sisters' needs; so, until we were given some land, there was no way we could start to work building a monastery. "Let's go ahead and get established in the state of Texas as a nonprofit organization anyway," I urged Madre Clara. That took two American citizens. I counted for one. Clifford Zarsky, who we had already by this time asked to be a member of our "board" and who was acting as our lawyer, became the second. We filed the voluminous papers.

Finally, a ray of hope glimmered. Madre Clara and I were invited to attend as guests the next federation meeting of the Poor Clares which would be held at the Minneapolis monastery. "We will ask them at this meeting," I told Madre Clara, "to accept us into the federation. If they do that, we can then discuss with them the building of the monastery in Corpus Christi." I knew that the Cuban sisters long-term chances of survival depended upon their being accepted with some status as members of the federation.

But we were met only with skepticism. We were not a canonically established monastery. We did not have approval from Rome. The Cuban sisters were a closely knit community with ancient customs from a strange culture, and this marked them as an easy target for misunderstanding. We were refused admission into the federation.

"I'll ask Father Dismas Bonner, the religious assistant for the federation, for a private meeting," I consoled Madre Clara. "Perhaps he can help us eventually to get accepted into the federation." But Father Bonner had been adamant. "Do *not* try to establish the community," he warned me. "Send the sisters, two by two, to other established Poor Clare monasteries who would have them." When Madre Clara and I had left the meeting with Father Bonner, we were deeply disappointed and discouraged.

"I have just spent the worst night of my life," Madre Clara said to me the next morning as we met to take our morning walk through the forest of Christmas trees grown by the Minneapolis monastery. "I also couldn't sleep," I confided. The truth was I had spent the entire night in prayer, not knowing which way to turn. Now as we walked in the snow among the rows of Christmas trees, Madre Clara was crying, and I was crying with her. Madre Clara kept saying, "How can I ever go home and tell the sisters we are going to be separated? They will grieve to death. After all they have been through and now this." I had never seen Madre Clara more dejected.

Suddenly we came upon a statue of the Sacred Heart of Jesus there in the middle of the forest. Together we knelt at the foot of this statue, weeping and praying. Madre Clara sobbed, "I can't believe God wants this." "Nor do I believe it, Little Madre. I will tell Father Dismas we will not be split up. Somehow, we will have a monastery."

That moment was a turning point. Then and there Madre Clara gave me permission, in her name, to do whatever I could to hold the Cuban sisters together as a community. At that moment I became a Cuban in heart, with the determination to let nothing stand in the way of what we both felt, after deep prayer, was clearly God's Holy Will for these beautiful, prayerful, united women. But our faith and love were yet to be tested. We returned home to face the future.

"Let's keep looking for land," Madre Clara urged me. So back to looking we went. One place we heard about was a beautiful place with a swimming pool and rustic home in a lovely residential district. But there was not enough ground and the home would have been very difficult to adapt for monastic living. Madre Clara and I decided against it. When Bishop Garriga heard about it, he was still trying to convince Madre Clara that it was good buy at $40,000. Again, she said, "No." The Bishop said to her, "Well, what do you want?" And she replied, "*Campo, campo, campo.*" Country land. Country land. Country land.

This conversation took place on February 11, the Feast of Our Lady of Lourdes. Was there any connection between that date and what occurred immediately thereafter? A realtor, Frank

Tompkins, said he knew about a piece of property that just might suit us. Out on Yorktown Boulevard in Flour Bluff there were over 200 acres for sale, and the owner would divide this acreage into twenty-acre lots. Madre Clara and I drove out and discovered that this site was exactly right. This land was perfect. Now all we had to do was to convince the Bishop.

I had been favored a few months before by a visit from my sister Mary. When Mary returned home, she called and asked, "How would you like to have a little monkey?" "Oh, great, that's swell," thinking she was only kidding. But, lo and behold, when my birthday came, I got a call from the airport. "Sister Bernadette?" "Yes?" "Would you please come down and pick up your monkey." "My *what?*" "Your monkey."

I had decided to name the monkey after Bishop Garriga and somehow Garriga got shortened to Gigi when we discovered she was a girl. I got a lot of amusement out of this little animal. In my room was a curtain that separated my bed from the desk. The curtain was stretched on a clothesline and Gigi would run along that clothesline and then hang by one foot. She would also watch me writing. Sometimes she would make one swoop down from the line, land on my desk, pick up a pencil, and then jump back up on the line again to chew on the pencil. After we found the land we wanted so much in Flour Bluff, I would tell Gigi over and over, "You're the Excellency's namesake. Can't you get him to give us the money?"

Then early one morning we got a message. Bishop Garriga had passed away in his sleep. Then a week later, another call came, this time from the chancery office. It was Monsignor Thompson. "Come down to the Chancery Office right away," Monsignor told us. "Right before Bishop Garriga passed away, he had signed eight checks. The bank will honor his signature. Bishop Garriga has given you a check to purchase that land you want in Flour Bluff."

With that check and a loan from the Catholic Women's Fraternal of Texas, we purchased the twenty acres and made plans to build the monastery. All this was expedited by Charles Kaler, a businessman who became my chief advisor for twenty-five years. Bishop Marx came out and blessed the land, although he would not walk through it because of rattlesnakes.

We also heard from our newly appointed Bishop Thomas Drury, who wrote to us from Rome. "You have permission to erect a monastery," Bishop Drury wrote us, "and when that is accomplished I will come and officiate at a Holy Mass. . . ."

Then Bishop Drury told us the almost unbelievable good news. I had urged Madre Clara to risk informing the Bishop about the plight of Caridad de Velasco. "For more than sixty years she has served this community," Madre had told him, "always waiting to become a Poor Clare nun." She had told him how Caridad came to be at the monastery, how she was always the first one at prayers, how she was never late, how for all her life it had been all or nothing for God. She told him how smart she was—a genius with memory, dates. How she knew the date every cow was born, the dates of every Pope's tenure of office, the specific circumstances of all matters around the monastery. How everyone was amazed at her keen mind and constant clarity.

"Please, Father," Madre Clara implored the Bishop, "when you go to Rome, ask for a special dispensation. I know it's never been done before, but surely these circumstances merit an exception. Ask if Caridad de Velasco can receive the rights of a full-fledged sister without having to be a postulant or a novice."

And now here was the letter with the Bishop's answer. I read further.

"I will come to officiate at a Holy Mass on the occasion of the Solemn Profession of Sister Caridad de Velasco."

The special dispensation had been granted! At age eighty-two, Caridad de Velasco would finally get to become what she had strived for since she was twenty years old. She would become, as a result of the appeal to the Vatican, Sister Caridad de Velasco, a choir sister, a Poor Clare nun. The news that we could build a monastery had been good, but this news about this rarest of rare exceptions for Sister Caridad was even more wonderful.

CHAPTER THIRTEEN

EARLY DAYS: LA FUNDACION

> May he give you your heart's desire
> and fulfill every one of your plans.
> May we ring out our joy at your victory
> and rejoice in the name of our God.
> May the Lord grant all your prayers.
> FROM PSALM 20

What a special relationship we felt with the land on Yorktown Boulevard.

From that first day when the realtor Frank Tompkins took us to see the acreage—"We're going to have to crawl through this barbed-wire fence to be able to walk the land," he told Madre Clara and me, and we bound our long habits as closely around our legs as possible to be able to crawl through the small opening he was able to make among the strands—we felt this was a hallowed place.

That first day we had walked down the cow path from the front of the lot, down to the middle of the land where there was a clearing of tall grass. Suddenly, as we walked along, we heard the rustling of wings and there, from one side of the clearing over to the other, flew a flock of at least twenty wild turkeys. We heard this "gobble-gobble-gobble-gobble" and then . . . swish . . . the beautiful birds flew by us. What an exciting moment! Mother and I looked at each other. Then Madre Clara said, "This is beautiful out here. This will be our land."

And, now, through the grace of God that had led Bishop Garriga to change his mind right before he died and write a check for us to use to buy the land, we owned this little piece of "el campo." The next step was to have plans drawn for the monastery and begin the work.

Leo Lopez, a talented architectural designer donated his time and prepared a set of plans. When Madre Clara and I went to his office to pick up the plans, however, we realized that somehow we had not communicated our needs clearly enough. The proposed floor plan was excellent for family living but not satisfactory at all for a monastery.

"Let's stop by the office supply store," I told Madre Clara on the way home. "I'll get some graph paper and see if I can design the monastery myself."

As soon as we got home, I set to work measuring the space needed for the refectory, chapel, sacristy, bedrooms, workshop, infirmary, and guest quarters. In about a week, I had designed the whole monastery, even to the measurements. Madre Clara and others helped me get down even the smallest specifics: how much room it would take up for each sister in chapel, the type of furniture we were going to put in there and in all the other rooms, the placement of the sacred statues. I drew everything to scale to allow for twenty sisters. When the plans were finished, Madre Clara and I then took my drawings back to Leo Lopez who prepared the blueprints. We were ready to build.

The date was December 8, the Feast of the Immaculate Conception of the Blessed Virgin Mary. This afternoon the first scoop of dirt would be bulldozed from the land!

"Let's go by Tom Graham's house first," I advised, "and watch him start that bulldozer up. Then we'll drive on to the land." The three of us must have been a sight standing there staring reverently at this big piece of machinery, waiting to hear the first loud, staccato bursts of the bulldozer motor. When we heard the sound, we clapped in joy. This moment was the beginning of the clearing of our land!

"Let's sing the Magnificat," someone suggested as we drove down Flour Bluff Drive. So with sounds that seemed to swell to crescendo inside our closed car, the three of us began to sing. . . .

> My soul proclaims the greatness of the Lord,
> my spirit rejoices in God my Savior
> for he has looked with favor on his lowly servant.
>
> From this day all generations will call me blessed:
> the Almighty has done great things for me,
> and holy is his Name.
>
> He has mercy on those who fear him
> in every generation.
> He has shown the strength of his arm,
> he has scattered the proud in their conceit.
>
> He has cast down the mighty from their thrones,
> and has lifted up the lowly.
>
> He has filled the hungry with good things,
> and the rich he has sent away empty.
>
> He has come to the help of his servant Israel
> for he has remembered his promise of mercy,
> the promise he made to our fathers,
> to Abraham and his children for ever.

This was the song the Poor Clares and I had sung in New Orleans the night I entered the cloistered monastery for the first time. This was the song that had been the Cuban nuns' anthem for hundreds of years in their ancient monastery in Havana. This was the song my sister, brother, and I had been taught so early by our mother to sing as a thanksgiving and adoration prayer.

When we arrived at the land, we had time to walk the stake line before the bulldozer came in. Then Tom arrived. "We will say the rosary," Madre Clara said to Sister Rosa and me, "while Tom starts clearing for the monastery." The three of us began to pray. When we finished and lifted our eyes to look around the land, we saw that the sun was almost setting. Oh, what a beautiful sight and what a beautiful sound. The tall grass swaying ever so slightly in the wind, the pink and purple and gold rays of sunlight fanning out on the horizon far to the west, and Tom Graham running that bulldozer back and forth as he cleared the land. "Glorious day!" we repeated again and again. "Deo gratias! Deo gratias! Thanks be to God."

"It is my theory," I told Madre Clara, "that the workmen will stay on the ball if we are out at the land nosying around every day." So while the building was going on, Madre Clara and I drove out as early in the morning as possible and stayed all day. "To be on the spot," Madre Clara explained to the other sisters who would be praying for us and for the workmen throughout the day. At lunchtime we would run in quickly to Van and Dottie Vermuelen's Drift Inn, pick up a hamburger, and hurry back out to the construction site. If we thought too many interesting things would be going on at the site, we would bring our lunch and forgo the fun of a lunchtime break with our new friends at the Drift Inn.

One day I pointed out to Madre Clara, "The workmen have a way of wasting things, like dropping nails and not bothering to pick them up. Let's you and me go around at the end of every day and pick up these nails. If we don't, they will just rust in the ground and then nobody can use them." Soon we had a big box of brand new nails which we knew we could use when the building was over.

Madre Clara caught on quickly. "Too much wood saw off . . ." Madre told me. "We save." She was right. When the workmen were putting on the roof, they would saw off the extra plywood overhang, leaving strips perhaps as wide as six inches. "Waste," Madre Clara said, "We save and later use." I agreed, and we began to collect these scrap pieces of lumber whenever the men were finished.

But now we needed somewhere to store all these finds. One day looking in the paper, I saw that someone was trying to sell a ten-car garage for $100. Madre Clara and I went to look. Yes, it would be perfect for our use. Oh, to be sure, the exterior looked terrible and all the ten garage stalls were open in the front, but the building would be perfect for our use. "Not only can we store the construction materials we have picked up," I told Madre Clara, "but we can get the building ready to store our ceramic molds when we move into the monastery." Seventy five dollars got the strong structure moved onto our land, and Madre Clara and I began immediately to renovate it.

Someone had given us a big load of composition tile, thinking we might be able to use it for the monastery. "We'll borrow a tile cutter," I told Madre Clara, "buy the proper kind of nails, and cover the entire garage with these shingles." It was hard work—the building must have been at least 100 feet long— but we persevered until we finished. "Job looks real professional," Jon Held, the contractor, told us when we finished. Madre Clara and I had to admit that we were pleased with our accomplishment.

One day we were sitting down in the garage eating our noon-time sandwiches when Jon Held came down to ask us something. Seeing suddenly all the nails and pieces of lumber we had salvaged, he said, "So, this is where all the supplies are going! What are you doing? Stealing from yourself?" We didn't even bother to explain that we had scavenged around picking up the discards because Jon was having such a big laugh.

We had the last laugh though. When Tom Graham told us that if we would get a herd of goats, the animals would eat all the underbrush on the backside of the land and clear it for us, we took his advice. "We will have to string a fence across the back of the property to contain the goats," I told Madre Clara, "but that and purchasing the goats will be much less expensive than paying someone to clear the land." So we got the goats. But then Madre Clara insisted, "We must build shed for goats. Too much for be out in weather." I also realized that a shed would be helpful at milking time. We were discovering a second valuable use for the goats—the Cuban sisters loved their milk. So every morning and every night I was now milking.

Using the discarded nails and the scrap lumber we had been

collecting, Madre Clara and I built a shed. Supplementing our stash of "loot," as Jon Held jokingly called it, we bought 4 by 4s to make a frame. Then we took all the trimmings we had gathered over the weeks and months and began at the bottom, overlapping the scraps that were about eight feet long and five or six inches wide, like shingles. When we finished, we painted the strips and had a terrific looking shed for the goats.

One morning shortly after Madre Clara and I had arrived at the construction site, Bill Johnston, a gentleman who lived down the road, came by and asked if we would like to have a little pup. "My English sheep dog has just had a litter," he explained, "fathered by nobody knows who. But the pups are cute." "Oh, we'll take one. Yes, we'll take one," I interrupted, forgetting this time even to look at Madre Clara much less ask her permission. But I needed not worry, for Madre Clara wanted the puppy, too. "We need sheep dog puppy," she responded. "Help me later herd our cows."

So Osita, who looked like a little cuddly bear, came to live in the garage on the construction site. "You stay right in this stall," I warned her, remembering how I had trained the puppies of my childhood. I did not want her to get in the way of the construction. "Shoo, get back in there," I'd say if the waddling ball of fur tried to follow Madre Clara and me out. And when we went home at night, I stringently instructed, "You stay right here after we leave; don't you go anywhere!" The little dog would obey. Every morning I'd hurry down to the garage as soon as we got to the site. There Osita would be, standing in the garage waiting on us. I would breathe a sigh of relief. Often I stood there for several minutes, my mind stirred to memories of the past. . . .

"Nina, whatever you pray for, if you have faith, God will give it to you," Mama kept reminding me as she prepared her seven-year-old for her First Holy Communion.

"You mean, anything?" I asked Mama, incredulously.

"Yes, if you sincerely and deeply believe that God will give it to you." Then she added, "You should pray for a spiritual gift and for a material gift."

So evening after evening, when I said my silent prayers at Mama's bedroom altar, my study for Holy Communion over for the day, I would ask God, "Please make me a good girl and please give me a puppy."

When the day of my First Holy Communion arrived, the family gathered back home afterwards to celebrate with festive meal. I got presents from many people, but the one I loved the most of all was a little curly white poodle puppy.

Then I'd think of other dogs we had. . . .

Pat, the airedale, who could not stand to hear a certain sound. I don't know how we discovered it, but if we kids said, "La-la-la-la-UUUUU," Pat would put back his head and howl, "AHUUUUUUUUUUUUUUUUA." Mama would reprimand us, "Don't do that to that poor dog; it hurts his ears." But every once in a while, when we didn't think she'd hear us, we'd taunt old Pat just to hear him howl.

Toots, the tiny toy poodle, that I used to clip to look like a lion. One summer we put Bruds under a big box out front by the sidewalk and put Toots on top of the box. "We have a talking dog," I would brag to the children who came along. "Do you want a bone?" I'd ask Toots, and Bruds would reply from inside the box, "Yeap," in what sounded half like a dog and half like a person. Then I'd ask Toots other questions which Bruds would answer. Finally, the kids caught on, but we had them fooled for a long time.

On occasion, the memories would bring pain. Standing there in the open door of the garage stall on the new monastery grounds, feeling my heart go out to Osita, this little bear of a puppy now in my charge, I would remember....

July 21, 1934. The day before I entered the Convent. Sixteen years old, alive with the thought that I was about to do something "big" for God, I surprised myself by feeling at that moment no sadness at leaving my family and home. As I bolted out the door to ride my bicycle to Bargaintown to the swimming hole for one last dip, Mama said, "You'd think you'd want to stay home with the family the last day." But off to swim I went.

After Mama and I had our picture made that night when supper was over and it was almost time for Papa and Mama to take me down to the convent in Pleasantville, I did have to say one good-bye. To my dog, Brat. Sis was upstairs sick in bed; I don't remember where Bruds was at the time. All I could think about was saying good-bye to my little companion.

Riding along that day on my bicycle, I had looked at someone's driveway that was filled with a lot of little pups. "Come on in and play with them," the man had said to me. Taller than a beagle but with those long ears and sad eyes, the little pups—there were fourteen in all—were black with brown feet and a little bit of brown on their bodies. "How would you like to have one?" the man asked me. Oh, it sounded too good to be true, but I responded, "I don't think my mother will let me have one." "If she doesn't," the man responded, "you can bring it back." So off I had gone with one of the smallest, a little male. "Mama, can I have him?" I kept asking, as Mama looked with disbelief from her bed upstairs where she was nursing a sore throat. "Well, get him off my rug; go ahead and take him outside." Oh, was I ever happy!

Brat was the most lovable dog in the whole world. He went everywhere I went. In fact, when we had a school picnic and went to an amusement park where they had a skating rink, I took Brat with me. I asked Sister Vincent Marie to hold the dog while I skated and he never took his eyes off me.

I would have him in the car with me down in Pleasantville when we would be waiting on Mama, and he'd be so cute with just his head sticking up, looking out—he was too short for passersby to see the rest of him—and people would stop and say, "What a cute little dog." I felt so proud.

Once I saw the man who gave me the puppy. "Is that the dog you got from me?" he asked me. "Yeap," I replied. "That dog is really being well taken care of," he answered. Again, I swelled with pride.

Yes, I had to say good-bye to Brat. He was the thing I most hated to leave. And in the weeks and months ahead, it would be Brat that I missed most of all.

One day Jon Held, down looking over our goat herd, suggested, "Why don't you also raise rabbits? I will buy all you can supply. There's nothing better than rabbit meat in chili." So up went the rabbit hutches. "Let's add guinea pigs," I urged Madre Clara. "I know we will be able to sell them to the pet stores in Corpus Christi." Soon guinea pig cages were a part of the growing animal quarters along the back fence.

Often I would experience déjà vu as I opened the rabbit hutch to place food pellets inside or when I picked up a baby guinea pig to move her. Perhaps the scene would be the day when, at the age of six, I inadvertently stepped on a mouse as I ran down to feed the rabbits in their hutch. "Oh, Dot," I called out to a friend I saw coming down the sidewalk, "come and see what I have found. This little mouse's stomach kept on quivering after I stepped on her. So I went into the house and got Mama's paring knife . . . and, look, what I found! A bunch of babies!"

Or checking on the guinea pigs, I'd remember the day walking home from school when I saw that the store around the corner had become empty. "Oh, if I could just rent that store, I would start a pet shop." I thought and thought of all the things I would put in it . . . the fifty turtles I already had, the tadpoles . . . and all my rats and guinea pigs. Then I thought of my classmate who also raised guinea pigs. "Would you like to go into business with me?" I asked him the next day at school. "We'll rent the empty store and have a pet shop." "That's a great idea," he had answered.

So the two of us went down, copied the telephone number off the store, called, and then went to the house where the owner lived. "Could we have the key to that store?" my friend and I asked. "We want to look it over. We may rent it. We want to start a pet shop." The owner, even if seeing two bright children before him, also knew we were only about eight years old. "I am sorry," he said definitively, "I am sorry, but it will not be possible for me to give you the key."

"You may end up having the equivalent of a pet shop yet," I laughed at myself as I stood looking at the rabbits and guinea pigs on the monastery grounds in Corpus Christi. Little did I know the extent to which those words would prove to be more and more prophetic.

CHAPTER FOURTEEN

ANIMAL LOVERS

You make springs gush forth in the valleys:
they flow in between the hills.
They give drink to all the beasts of the fields;
the wild-asses quench their thirst.
On their banks dwell the birds of heaven;
from the branches they sing their song.

From your dwelling you water the hills;
earth drinks its fill of your gift.
You make the grass grow for the cattle
and the plants to serve man's needs.
FROM PSALM 104

We soon added a donkey. Madre Clara, Sisters Martha, Encarnacion, Ascension, Rosa, and I were on our way home from spending time on the land, sitting under the trees in our lawn chairs, eating crackers and drinking Sprite. Before we left, we had taken some pictures. Now as we drove along in the car, I saw the sign, "Ponies for Sale." "Let's stop," I entreated Madre Clara. She agreed, and the sisters got out to check on the situation. "One left," Sister Rosa said when they came back to the car. "We don't have time today to check on it," Madre Clara said, "but go tell the man we'll come back out on Sunday."

We stopped on Sunday and saw the pony. She was eleven months old and very tame. Madre Clara seemed pleased that the pony followed her when she walked away. We put $10 down and promised to come get the pony when we had water on the land. We paid the $75 balance after the vet came out and checked on her and said she was a bargain. Then we put the pony in the back seat of the car—two nuns, one man, one pony in a Chevy—where she rode standing up and not a bit nervous as we made our way to her new home.

I picked up a halter for Peanut at Sears where I discovered that the salesman used to ride the horses at Sauners Stables in my hometown of Northfield. Imagine!

Then I stopped at Horseshoe Inn where I got two salt blocks for Peanut and looked through their catalog for a bridle and bit.

In passing, I spotted a darling two-wheel pony cart for only $180. But then the salesman said he knew where we could get a four-wheeler for $50. Madre Clara and I rushed over to the place where the cart was for sale and discovered that it was, indeed, a bargain. We bought the cart and bridle for $60, and Bertha Eiermann brought it home in her station wagon with Madre Clara holding the shafts from the front seat. "This cart is very strong," I told Madre Clara as we were loading it. "It will be able to hold three little kids, two medium-sized kids, or one of us "big kids." I was already imagining rides around the monastery grounds like the rides we kids would take with Mama or Sis driving. Ah, what sweet memories . . . Mama waiting on me every day when I was in the first, second, and third grades. I was the envy of all my classmates when I climbed up into that pony cart to go home with Mama.

On a very rainy day, six months after the construction began, we moved into the monastery. The first thing to be loaded on Didear's moving van was the large statue of Our Lady of Caridad which the sisters had somehow managed to bring over with them when they escaped to America. To fulfill a promise the sisters had made to the Blessed Virgin Mary, we gave the statue a place of honor in the new chapel with its heavenly blue walls and its clean white ceiling.

In August I had a phone call from Bishop Drury. "Sister, I am going to proclaim the second week in August as 'Poor Clare Week' and I want you to have open house every afternoon and evening. I am sending out a photographer, and you get the situation organized."

And organize we did. I had met Lois Wilder through her husband who was trying to sell us grave lots. He ended up buying a book from me. Lois had come, then, to the monastery to see the person who had pulled such a fast one on her husband. We became immediate friends. Now she was taking on the task of organizing the whole week. We had persons stationed at every location in and around the monastery. Over three thousand people attended the open house and enjoyed refreshments and entertainment in the refectory.

Mary, my sister, came for the festivities. "What in the world are you doing with this cage of twenty-four parakeets?" she asked upon her arrival. "Don't you have enough to do taking care of this big place?" "We're starting to breed parakeets for a local pet shop," I answered. "Don't the little things remind you of the beautiful birds Mama used to keep in that wicker cage in the sun room?"

When Mary returned to New Jersey, she happened upon a pet shop in a newly opened mall. "Where do you buy your birds?" she asked the owner. When he told her, she advised, "You should get them from the Monastery Bird Ranch in Texas. They raise gorgeous parakeets." Thus came into being the Monastery Bird Ranch. Within that same week I was on the telephone, engaged in a conference call, with the owner of the Docktor Pet Center and their bird buyer. By the time the conversation was over, I had agreed to ship parakeets to their stores all over the country. And we had only twenty-four parakeets in the whole place.

The first order from Docktor Pet Center could have wiped us out. But, undaunted, Sister Rosa and I traveled to many little towns and out-of-the-way places in the Rio Grande Valley to hunt for birds. Our first stop would always be the feed store where we would ask to whom they sold bird seed. From there we would head out to find the people who had bought the seeds. We purchased many birds at the aviaries we found, but we also observed. How were the aviaries set up? What equipment did they use? What did the best aviaries have that the others didn't?

The local bank got to know me. We did not, of course, have any capital. So on Friday the bank would give me a loan of $1,000. On Saturday I would buy sometimes up to five hundred parakeets. By this time many of the breeders were coming to us instead of our having to go to them. I had deliberately paid the breeders a slightly higher price than they had been receiving from the representatives of the large bird companies so that they increasingly transferred their loyalty to us. On Monday morning we would take the whole lot of birds to the airport to be shipped all over the United States. I would then go to the bank and repay the loan until the next Friday.

Marie Colson, down the road from us, got me interested in Himalayan cats. Before we knew it, the aviary had also become a beautiful cattery with forty-eight individual eye-level cages for queens about to kitten. We called it the maternity ward. Eventually, we had about seventy-eight queens and a dozen toms. The cats had large areas outdoors for roaming and playing and could come inside to eat and sleep. We also built a special grooming area. We shipped kittens to the same customers that we previously had shipped birds. Since all the cats were registered, we spent hours and hours on bookwork to supply to our customers complete pedigrees with several generations.

Then one day I got the bright idea that we should expand further and raise parrots. "We are so close to South America," I told Madre Clara. "I know we can find great sources for parrots. We can begin with a few and see what happens." The brightly colored birds were as big a hit with our customers as the cats and parakeets, so soon the aviary-cattery became home to dozens and dozens of parrots.

"I hear you're looking for yellow-headed parrots," the voice said when I picked up the telephone. "Yes," I replied. "Would you like to buy about a hundred?" A hundred? I could not believe our good fortune; to be able to buy that many healthy parrots in one fell swoop was unheard of. "Certainly, I'd like to see them," I answered. "Where can I find you?" "Brownsville," the man answered and then gave me the address where he was staying.

The next day Sister Rosa and I sped on our way to Brownsville in our new leased-to-buy station wagon. Little did we suspect that we were headed for a dramatic encounter with the underworld!

We had no trouble finding the man's house in Brownsville, but we were rather surprised to learn that he had to take us across the Mexican border into Matamoros to show us the parrots. This was a turn of events he hadn't mentioned on the telephone. "We've come this far," I told Sister Rosa. "It won't hurt us to drive fifteen or twenty more miles to check out the parrots." So the gentleman got into our station wagon and we drove to Matamoros.

We could hardly believe our eyes but there they were—one hundred perfectly adorable squawking, hand-fed yellow-headed babies. A literal gold mine. But then came the first hitch: We could not load the baby parrots into our station wagon in the heat and would have to wait until sundown. "No problem," Sister Rosa and I responded. "We'll just wait in the shade on the Brownsville side of the bridge for the birds to be brought through customs." The second hitch was that the man would accept only cash. Looking at the birds and realizing how much we would be able to sell them for to the pet stores we supplied all over America, I quickly determined we could solve that situation. "No problem," I told the man. "We'll go to a bank in Brownsville and get the money this afternoon."

So back across the bridge into Brownsville we went. The beautiful hand-tooled leather saddle I had bought while in Matamoros sailed through customs with no trouble. The nice man who was with us had chatted in a most friendly manner with the guard at the border. Apparently they were well acquainted. "How lucky for us," I thought. I knew that parrots were difficult to find but almost impossible to get across the border.

And now Sister Rosa and I were sitting in the station wagon under the shade of a palm tree in Brownsville, waiting for the parrots. Blissfully, we ate bananas and watched the blazing sun sizzle out of sight. Uneasiness began to creep over us as the sun disappeared. Eventually we decided that although the prospects of making a lot of money on those parrots looked mighty good, the circumstances of having to travel at night was not a welcome situation. "Let's give up this long wait," I told Sister Rosa, as I started the motor. But just as we started to pull out, a car drove up beside us and the nice man appeared saying the parrots had just arrived.

Then everything happened very quickly. The man and his helpers shoved the enormous crate of squawking parrots into our new station wagon. I quickly retrieved the money Sister Rosa was holding and without question the man took the wad of money, closed the back of the station wagon, and we all took off.

Breathing a sigh of relief, I headed for the gas station. But we had gone only about a half a block when suddenly we were surrounded. One car pulled directly in front of us, one blocked us from the back, and one edged me over to the curb. Out jumped a man who shoved a card under my nose and read me my rights.

"Say, what is this?" I demanded.

"You are under arrest," the man snapped back, "for carrying contraband."

"I'm carrying nothing of the sort," I retorted. "These parrots were purchased right here in Brownsville."

"Sorry," the man replied, "you must come to customs." Then looking at Sister Rosa he said, "You get out and get in that other car. We'll drive you to customs, too."

"I no leave," a very determined Sister Rosa answered as she refused to move. "I stay right here with Madre Bernadita!"

"Sorry, she won't budge," I told the man, "and besides that I won't let her."

"Then follow us," was his answer.

The procession of cars headed back to customs at the border. Once within the building they separated Sister Rosa and me and began to question us. Of course, we both had the same story. As I sat there being interrogated, I could smell feathers burning. An officer at the border appeared saying, "We had to destroy the parrots. That's the law. And your station wagon has been impounded. Come back tomorrow to talk to the officer in charge of that department."

"But we are from Corpus Christi," I spurted, not believing what I was hearing.

"There is a hotel just down the street," was his curt reply. "It's within walking distance."

So together Sister Rosa and I headed to the hotel, which was the only thing we knew at that moment to do. Safe in the hotel, the door to our room securely locked, I had a sudden frightening realization. "Oh, Sister Rosa," I said, trying to keep my voice as calm as possible, "do you realize that the money you were carrying was only $500 and that the other $500 was in the glove box and in my purse. We didn't give the man who brought the parrots all his money." "That man will come back and get us," Sister Rosa lamented. "I know he will get us." I tried to reassure her, but I wondered, too, how long it would be before there was more trouble.

We had called back to the monastery, telling them the car had broken down and we would be home tomorrow, so no one there was worried. But Sister Rosa and I were up and down all night, looking through the slit between the curtains to see whom we could see, jumping every time the air conditioner went on and off. Finally, it was morning and we could walk back down to customs to get this nightmare straightened out.

"Sorry," the man said when we got there. "You will have to wait to come back Monday. The man you have to see is off for the weekend."

"We'll call someone in Corpus Christi to come get us," I told Sister Rosa, but before I could make the arrangements, a man came into the room and said I was wanted on the telephone. It was our lawyer, Richard Hatch. During the night at the hotel I had called and confided in Bertha Eiermann about our escapade at the border, and her husband, Chester, had quickly called our lawyer.

"Where are you now?" Richard asked.

"We just walked into the customs office downstairs."

"Go immediately to the office of that official," Richard instructed me. "I want you to be there now when I phone him."

So up we went. By the time we reached his office, the man was on the phone so we just sat down. I noticed that he was not talking and his face was very red. In a moment he held out the receiver and said, "Someone wants to talk to you."

Of course, it was Richard Hatch. This is what I heard: "Listen carefully. I do not want you to engage in conversation with that man. When you hang up the phone, just say to him, 'Good-bye, sir!' turn, and leave immediately. Go back to that hotel and wait for me to call you there."

The man never said a word to us, let alone try to detain us, as I put the phone back on the hook, said "Good-bye, sir," and motioned to Sister Rosa that we were leaving. Back in the hotel we awaited the mysterious call.

Talk about anticipation. Sister Rosa and I kept trying to come up with ideas about what might be happening. Finally the telephone rang. It was Richard with some startling news. "You have been set up," he told us. "That whole thing was planned by a bunch of criminals. The officers at the border had no right to detain you nor to confiscate your car. You cannot get your car back right now, so come on back to Corpus Christi with your friends. I'll work out the details of retrieving the car later."

After several months had passed, we received a letter from Governor John Connelly expressing regret for the loss of the parrots and informing us we could pick up our station wagon in Brownsville. Meanwhile, because of the lease-to-buy car agreement, I had paid off the impounded car and bought another make from a different dealer. So we just sold our impounded station wagon when it was returned. "Just as well," I told Sister Rosa in private. "I don't want anything around here that would always remind us of that terrible experience." Needless to say, I had learned a most valuable, if painful, lesson.

By now we had fazed out the goats since their job of clearing out the underbrush had been completed and had bought milk cows. I milked the first year until my thumbs wore out; Madre Clara milked the second year until her thumbs wore out. Then we procured a small portable milking machine which Madre Clara loved to operate. So we always had fresh milk and sometimes even butter or cheese.

We bought donkeys to go with the pony. When Cola had a baby we named her Coke, and I even slipped the newborn donkey into

Sis' bed when she was there visiting. Later we replaced the donkeys with horses. It was fun to ride around through the wooded area on our twenty acres. When my brother or sister or cousin Charles came, we rode all over the land, reminiscing about the horse rides we loved so much when we were children.

We had chickens and roosters on the grounds, also. I could almost never pass the chicken pens without getting lost in reverie.

"Nina . . . tell me you didn't . . . tell me you didn't put that banty rooster on a leash and walk it all up and down the Atlantic City boardwalk!" Mama was appalled. "Yes, ma'am, that's what she did," Bruds said, even before I started to think up a justifying answer.

"Where are you going with that rooster in your bicycle basket?"

"Over to the post office. Here, read this letter."

Mother Clarita had called a few days before. "Nina," she said, "I was just wondering if you are still thinking about being a sister. There are three girls going to enter the convent next month in Allegany. I'll be taking them up. I just wondered if you were still interested."

"I'm interested," I told Mother Clarita.

"You must get permission from your parents and then write a letter to Reverend Mother Dominica and tell her you would like to be a sister and that you have your parents' permission."

Until this moment, however, I had not said one word to Mama or to Papa. I would just let her read the letter.

"You're not good enough to be a sister," Mama said when she looked up from reading. "No, I know I'm not good enough, but I'll get to be while I'm there," I responded.

"Well, if that is what you want, go ahead," had been Mama's reply. And I had jumped back on my bicycle where the rooster was still sitting in the basket and pedaled away to go first to the post office and then on to Bargaintown to go swimming.

Papa had reacted differently. After the letter had come back from Mother Clarita telling me I had been accepted and after

Mama had already bought the big black trunk and many of the items the convent requested that I bring with me, Papa called me into his room. "Come in," he said, "I want to talk to you for a few minutes." It was something very unusual to be called to Papa's room to discuss anything, so I knew what this was going to be about. I was embarrassed that I hadn't said anything to Papa before, so I picked up the big black umbrella Mama had just ordered for me and carried with me into the bedroom so I would have something to do with my hands.

"Do you know what you are doing, Nina?" were Papa's first words. "At the age of sixteen, you are giving up a family, a home, and children. Do you really know what you are doing?" "Oh, yes, Papa," I responded. "I understand and I want to do this," all the while that I was opening and closing that big black umbrella, open and close, open and close, open and close.

Finally, Papa said, "Well, if that's what you want and you're sure that is what you want, you can go and try it." Of course, at the time I had no idea how my going might affect those around me. All I could think of was myself and this big deal of going away to the convent and having a trunk and all the things that I got to carry with me.

The little rooster also went with me when Bruds and I went over to tell Father McCallion that I was going into the convent. I took the rooster all the way into the vestibule of Father's office and left him there while Father, Bruds, and I were talking.

Not all the animals on the land were welcome. Rattlesnakes were a constant threat. On the seventeenth of every month we had a mass offered in honor of St. Patrick, praying that no sister would ever be harmed. In all the twenty-five years we lived there, no sister ever received a bite. We had snakebitten horses, we had snakebitten guinea pigs, and we had snakebitten dogs. But we never had a single snakebitten nun—although we certainly had some close calls.

One day I was walking down the dirt road toward the garage where we had our ceramic shop in the beginning. A huge rattler shot right across in front of me within striking distance. The snake and I both froze on the spot. He coiled and began to try to locate me. I stood very still so he was confused. Then he began to rattle. It sounded like a thousand castanets. The sound was so loud that a worker down at the garage heard it and came running with his gun. The snake's whole body had coiled for a spring yet his head never moved from the spot. I gathered up all my energy and sprang directly backwards. With that the snake slithered off the road into the grass. I yelled and pointed out the snake to the worker who by this time was beside me with his gun ready to fire. That was, thank God, the end of Mr. Snake.

From that time forward every nun wore a whistle. We established a rule: If you come across a rattlesnake, do not leave. Stay in the vicinity, keep your eye on the snake, and blow the whistle. This way, we always knew that when we blew the whistle, someone would pick up a hoe or a shovel and come running. Madre Clara became expert with the hoe; she loved to sneak up behind the snakes and whack them. Her behavior worried me but she was careful and always got her snake. I kept my 410 and ammunition handy at all times, and this is what I used if I were the one who heard the whistle.

On another occasion we had just built a new cattery. The sisters had all just been down to see the facility for the first time. I had driven some of the sisters down since it was too much of a walk for them and had just returned to close up after the tour. Suddenly, as I reached up in the corner of the little porch to turn off a light, I heard a rustling that sounded like plastic rattling. "There's no plastic around here," I said to myself. I looked and right there, extremely close to me, coiled up, was a rattlesnake. It had to be Divine Providence that he had not struck. I retreated slowly and blew my whistle. Sister Rosa came running. "Go get my gun," I yelled, and the sister ran to the monastery as fast as she could. "I must keep dancing around to keep the snake up on that cement slab," I instructed myself. So dance around I did for what seemed like an eternity until Sister Rosa came back with my gun. "Where's the ammunition?" I asked her, excitedly. "I thought it was in the gun," she replied. "No, it isn't," I answered. "You'll have to go back to get the bullets, and please hurry." So off Sister Rosa went again as fast as her short legs would take her. In the meanwhile, I still danced around and watched the snake while a bunch of little kittens were looking out a picture window in the cattery, right about floor level. They were also watching the snake.

"Go closer," Sister Rosa kept repeating when she brought me the shot and I had loaded the gun. "No," I told her, "I don't want that shot to ricochet on the cement." I gave one shot and the snake went flying right straight up into the air and blood splattered all over the wall. The rattle on that snake was so big that I decided to embed it in some liquid plastic we had in the craft area so we would have evidence of God's protection for posterity.

Once the mailman discovered a big rattlesnake in the dog kennel right by the mailbox at the front of the house. He rang the doorbell to alert the sisters and then picked up a big stone, threw it over the fence so that it landed on the snake and pinned it down. The sisters were able then to come out and kill it. "At least five feet long," they reported to me when I called from Brenham. "Very thick around . . . with a huge rattle."

During these formative days for the Cuban foundation, my personal responsibilities were tremendous. Every decision, every action fell on my shoulders. I necessarily had to be at the service of everyone at all times. Occasionally, the load would get very hard. I can recall one day in particular. I was down taking care of the birds, which of all the animals we raised were the most difficult for me because I was quite alone in that venture. It was a very, very hot day and I was standing in the middle of the aviary sweeping up feathers, my ears bombarded by the chattering of hundreds of parakeets. I can remember just standing still and feeling big tears rolling down my cheeks. This was the first time I had ever cried over the difficulty of my work.

I thought to myself, "What have I gotten all of us into? God give me the grace to continue this." I knew we had to have the money to pay off the loan we had taken out to build the monastery, and I knew God wanted this Cuban sisters' monastery. But today it seemed that I could not find the strength to go on. Then as if it were just yesterday that I had sat in the classroom at St. Pete's elementary school, I heard Father McCallion's favorite motto: *"I am but one, but I am one. I cannot alone do it all, but I can do something. What I can do I ought to do. What I ought to do, God helping me, I will do."* The words gave me at that moment an almost miraculous strength as I wiped the sweat off my brow and continued my sweeping.

CHAPTER FIFTEEN

TOUGH TIMES AND FUN TIMES

> I will praise you, Lord, with all my heart;
> I will recount all your wonders.
> I will rejoice in you and be glad,
> and sing psalms to your name, O Most High.
> FROM PSALM 9

The monastery was situated not far from the Gulf of Mexico. This meant that often the elements were cruel to us. The worst we feared was the hurricanes. Many of these storms we were able to ride out battened down in the monastery, but when Beulah came through we knew we would have to leave.

Father Claude Valentine, who had come out to offer mass, had given me definite instructions. "When the National Guards come to tell you that you have to evacuate," he counseled, "give Communion to the sisters and consume the extra Hosts yourself. Move everything possible away from the windows and as high off the floor as possible. Then get over the bridge and into the main part of the city."

When the word came later that afternoon, we sent some of the sisters to Spohn Hospital for refuge and the rest of us took shelter with Laura Rodriguez in her stone house in town. Laura had given English lessons to the sisters when we lived in the house on Third Street, so we all knew her well.

We slept as best we could and anywhere we could. My bed was on the floor under the dining room table. The hurricane hit full force about one or two o'clock in the morning, and I had my little transistor radio up to my ear to hear any reports. The last news broadcast before the hurricane stopped every form of communication came from a reporter who was driving through different parts of the city. I will never forget that final broadcast: "We are on Flour Bluff Drive but can get no further. The gulf water has risen past Waldron Road so we must turn back." My heart was in my throat. That report had been given from a location only one block from the monastery.

Sweet dreams? I visualized the monastery submerged in seawater. It was likely that we had lost everything. I felt desolate and distraught. Then I felt the urge to pray. "Oh, God," I began,

"I give back to you now everything You have so graciously bestowed upon us. I give You the buildings, the chickens, the guinea pigs, the dogs, the donkeys, the horses, the birds, the cats. I give you back all the furniture. God, I give you back everything." With that prayer I fell sound asleep on the floor, warm and dry and happy that we were all alive.

The next morning we awoke only to the sound of rain. There was no raging wind, just rain. Pat and Cucca DeLeon, the caretakers who lived in a trailer on the front of the monastery property, were the first to venture out to Flour Bluff. Pat was able to get through on the phone to us. What a surprise. He was calling from the monastery. The announcer had mistaken the heavy rain for seawater. It was true that we were flooded but by the rain, not by the sea.

A few of us headed out for the monastery after the telephone call. The heavy rain was still falling, rain that would not let up for over a week. Devastation was everywhere. When we got to the monastery, our entire backyard was under water. I myself had planted two thousand feet of vegetables, corn, and beans, while other sisters had tilled okra, tomatoes, and watermelon and had raised beautiful gladioli for our altar. Now I looked upon a lake. Peering down into the water I could see stalks of corn about a foot high looking now like underwater vegetation.

And there were frogs everywhere. The whole place was a croaking mess. John Decker and I went to the chicken house. There were chickens everywhere. Floating, dead chickens. Two or three were perched on a roost out of the water, but all the rest had drowned and were already beginning to puff up.

"What in the world will we do with these dead chickens?" I asked John Decker. "There is no way we can get down to the dump grounds which are all under water and muddy anyway." "Well, let's call the police and ask them what to do," John suggested. We went into the monastery and I made the call.

"Sir," I asked the gentleman who answered, "what do we do with seventy-five dead chickens? We're going to put them into burlap sacks."

"Well," came the slow reply, "you don't have your name on the sacks, do you?"

"No, sir, of course not," I hastily replied, not mentioning that we didn't even have any sacks as of yet.

"Well, dump them anywhere you can," the voice instructed.

John gathered up all the smelly dead chickens and we put them in sacks—without our names on them, of course. Then off he went to leave them wherever he could. I began to survey the damage. The chicken house, of course, was completely destroyed. Sixteen windows in the monastery were broken. The whole chapel was flooded knee deep with water. Broken glass lay everywhere. We had no electricity, and the commodes would not flush. "The sisters cannot come back out here to this mess," I told myself. "We'll have to find temporary quarters until a few of us can get things cleaned up."

For the next ten days most of the sisters took shelter at the Incarnate Word Convent or at the Spohn Hospital. Four of us returned to the monastery to do the cleanup work. During the day we cleaned up broken glass, dragged out soaking wet bed mattresses, rehung venetian blinds that had swung around hitting furniture, and took water out of the chapel bucketful by bucketful—a job that seemed to take us ages to complete. At night we huddled together in the guest room, trying to feel safe enough to rest.

One night we heard a noise out front. The four of us went running into the guest room. I peeked through the blinds. Two men were standing there.

"What do you want?" I called out.

"We're officers, and we've come to see if we can be of any help," they replied.

I could see that they had on ordinary clothes so I said, "If you're officers, how come you don't have on your uniforms?"

"Oh, we're okay," they answered. "We're just helping out off time here," and they took out their wallets, held them up, and turned their flashlights on to show identification. Of course, I could not see . . . and I was praying for guidance all the time. Suddenly I had an inspiration.

"Sirs, since you are officers, maybe you could answer something that I've been wondering about for a long time."

"Oh, sure," they responded. "What is it?"

"Well, I know that if I should shoot intruders in my house I would not be responsible but what about if they are outdoors?" Then I added, "Oh, well, I guess it really wouldn't matter because I'd do it if I had to anyway . . . and I'm a crack shot."

"She surely is a crack shot," one of the sisters piped in. "Why not long ago she shot and killed two rabbits in the dark just by seeing their white tails."

"Well, if you need anything or want anything, just take your flashlight and make big circles like this," one of the men said, making a motion with his flashlight. "We'll come and help you if you need any help. So long." Then they scurried off.

Later on when we found out that there had been many looters in the area after the hurricane, we had no doubt that these men had planned to break into the monastery and instead had been surprised by finding us there. I thanked God repeatedly for protecting us and St. Clare for helping me know what to do in that moment of terror. For all the time I had been standing there with my gun ready to point toward the intruders, not knowing what the next moment was going to bring, I had remembered the promise of our Lord to St. Clare in one of her moments of danger and adversity: "Because of your love, I will watch over you and them always." This promise of protection, which St. Clare had received as she stood holding the monstrance in which the Sacred Host was kept while the invading bloodthirsty Saracen army were even then at the convent wall, had greatly sustained me.

The rain continued for days after the hurricane, long after we had returned the interior of the monastery to some order. Our chapel windows were very large, reaching from floor to ceiling, and it was easy to see that the grass outside the windows had become one big pond, with grass poking up and croaking frogs taking over. As I stood there listening to the frogs, I felt a rush of thanksgiving for both the tough times and the fun times. I realized that the combination of both is how we grow in the love of God. "God is ever present and blesses our efforts," I said to myself. And hearing again the concert of croaking frogs outside the chapel window, I added, with a smile to myself, "Why, *all* His creatures praise Him!"

Back in my room, I tried to capture those moments of thanksgiving and praise by writing. . . .

Frog Dialogue

Glass-like ripples
Wind-whipped puddle
Swollen to pond-size
Where grass blades
Poking up their claim
Reach above the mirrored surface.

All this right beside
Low chapel windows
Which reach to the floor
And divide our choirs
Where dry-singing nuns compete
with rain-wet dialoguing frogs.

We secretly laugh at their croaking
And smuggly raise our voices
Alike in praise of Him,
Creator of both nuns and frogs.
The competition is good.
Frog dialoguing will end but
Our singing is eternal!

After the rain subsided and the puddles turned into soft dirt again, we tried to get everything back to normal as soon as possible. Madre Clara and Sister Magdalena were anxious to buy more chickens to get our freezer full of broilers and the refrigerator full of good fresh eggs once again. With a small business loan of $3,000, we built a new chicken house, this time of cinder blocks instead of wood.

We began also to plant. Earlier we had purchased several fifty-pound bags of different seeds in order to provide our parakeets a variety of food. The adventure was a flop, for the birds would not touch the varieties we had bought. "We can't waste all these seeds," I said to the sisters. I had read somewhere that rape seed would produce edible leaves that were really a delicacy, so I planted the seed in long rows. The plants bore luxuriant leaves which, when cut, produced more and more. We had to eat rape leaves for a long, long time.

The hemp seed were a different story. Since the label on the bag said it was sterile seed, I merely scattered the seeds around, hoping the wild birds would profit by it. Then one day Lois Wilder came to visit the aviary. Walking down the narrow cement path, she noticed the neat rows of rape leaves. But before she could ask what they were, she gasped and exclaimed in a shocked voice, "Sister Bernadette! What are you growing over there?" She pointed to where I had scattered helter-skelter the hemp seed.

"Oh, that was supposed to be sterile seed," I explained, "but some of it has produced those plants with beautiful leaves. Aren't they attractive?"

"Attractive, indeed," she replied in a hurry. "That crop will soon be attracting the cops! Don't you know what it is?"

"Must be hemp plants," I responded. "What's so terrible about that?"

"*That*," she replied with great emphasis, "is none other than *marijuana!*"

I immediately began to clear out that garden.

I also learned something about this time concerning bay leaves. Sister Martha sent me to the grocery store to buy a pound of laurel. I paid no attention as the grocer weighed and sacked the herb, but when I got home I discovered that a pound of laurel was enough to last us for years. I also discovered that laurel was the same as bay leaves, and if there were anything we had on the monastery grounds that was plentiful, it was bay trees. So I promptly set to gathering bay leaves, drying them, and packaging them to sell around town and to our visitors.

Not only did we have an abundance of bay trees, but the monastery grounds were also covered with tall, wild grasses. This grass was the occasion for another experience similar to the night I faced the looters after the hurricane, when St. Clare seemed so close to me. "Fire . . . fire . . . fire . . . ," several of the sisters began proclaiming at once as they spotted a hugh fire which had started right across the road from the monastery. Welders' torches had sparked a brush fire which, because everything was extremely dry, was raging within minutes. As soon as I got to the front of the house, I knew the monastery was in great danger.

"What can we do? What can we do?" the sisters were crying and gesticulating frantically. I knew we could get out ourselves, but what about the cattery? There was no way, given how rapidly the flames were moving, that we would have time to gather the 200 cats and kittens, put them in cages, and get them off the grounds to safety. The situation was desperate.

Suddenly the story of St. Clare and the invading Saracens flashed into my mind again. . . .

Even as the bloodthirsty soldiers were climbing over the convent walls and banging on the convent door, St. Clare, rising from her sickbed, asked for the monstrance in which the Sacred Host was kept to be brought to her. This monstrance she held up in the large window opening for the infidel Saracens to see. It was then that a voice as sweet as a child's spoke from the monstrance to say, "Because of your love, I will watch over you and them always." Shortly there-after, the Saracens left the convent grounds and departed from the countryside without further invasion.

I was inspired by the memory.

"I will get the monstrance from the tabernacle," I told the concerned sisters. "You stand with me at the front entrance praying while I hold the Blessed Sacrament on high." This we did, even as the fire raged and jumped the road to our side of the fence. "Keep praying," I urged the sisters, as the fire, flames shooting up sky high, came closer and closer to the monastery. "I'll keep the Blessed Sacrament held high."

Suddenly, as we stood there praying and watching, the fire just burnt out immediately upon reaching the fence to the monastery. Not even a spark was flying. With prayers of thanksgiving the sisters hugged each other as I walked back in praise and adoration to replace the monstrance in the tabernacle. The fire was a topic of conversation for the rest of the afternoon and evening, however, for it continued to burn uncontrollably, having jumped another road and burning everything in sight. Firefighters were unable to extinguish the fire until late in the evening.

As time went on, the Cuban culture to which the sisters belonged and the American culture to which I belonged occasionally would scrape against each other, if not downright clash. Take, for instance, Cuban food.

The last straw was the evening I came home late for dinner because I had been taking care of business downtown. I went to the steam table where they kept food hot for latecomers. The steam table had about eight sections, three of which held hot foods. I picked up the first lid and there was the expected rice. "Okay," I thought, "I will see what is in the other two sections before I begin to dole out my dinner." I picked up the second lid and peered in. Beans, as I expected. Well, I had one more hope. I picked up the third lid and what did I find? Beans and rice mixed, leftovers from yesterday.

"I'm going to take over the cooking," I told Sister Martha and Madre Clara. "That way you'll be able to taste some new recipes." But I might as well have saved my energies, for nothing could ever please the sisters as much as their beloved beans and rice, which they continued to request daily.

And about physical exercise we also had cross-cultural differences. No matter what I tried, I could not get the Cuban sisters to exercise. I knew that pushing a broom and scrubbing a floor were not the only type of exercise that should be found in a monastery. Cloistered nuns live in very close proximity to one another, rarely coming into contact with outsiders. We must, therefore, find recreational and exercising facilities within a very limited area. This was the reason we had acquired some riding horses and why I worked so hard to help us get a pool.

We had two long buildings side by side. A pool forty-feet long but only twelve feet wide was installed between the two buildings, and a tall cedar fence closed the area off at both ends. One could enter the area directly through a rest room in one of the buildings.

My idea was that I would be able to get some of the older sisters to use the pool. I knew the hydrotherapy would do them a world of good. But no go; not a single Cuban sister would use it.

One day Bishop Drury visited us. Madre Clara and I were showing him some of the buildings. I will never forget Madre Clara and her consternation when we took the Bishop through the little bathroom and out into the pool area. He hadn't known he was going to be seeing a pool, so he was very much surprised. Madre Clara ran ahead of him and stood there looking at the Bishop, scared to death.

The Bishop took one look at the pool and said, "Isn't that great! How long have you had that?"

"A couple of months," I replied.

"Oh, I bet the sisters enjoy that!" the Bishop replied. Then addressing Madre Clara, he asked, "How do you like it?"

She exclaimed in a hurry, "Oh, no, no, no!"

"What's the matter; doesn't she go in?" the Bishop asked me.

"Your Excellency," I answered, "none of the Cuban sisters will go near it."

"What's the matter?" he asked. "Won't they let their hair down?"

If the Bishop had only known how much it bothered me that I was unable to interest the sisters in matters of exercise and play which it seemed to me they so much needed to balance off their heavy work.

We did have many good times at the monastery, however, which helped to lighten up our existence. Christmas was especially a time of joy and fun. Each year we had a live nativity scene for several days before Christmas. As Franciscan nuns, we wanted to imitate St. Francis' love for the infant Christ and felt that having a nativity scene was one way of doing so. "Remember how St. Francis brought live animals to the cave in Greccio during a Christmas midnight Mass?" we would ask one another. "Yes," someone would add, "the animals helped bring to the people a vivid conception of Christ's birth." Another sister would say, "The infant Christ is born again at Christmas in the hearts of men. Didn't St. Francis remind us that God Himself came to earth as a baby so that we would have someone to care for?"

God, a helpless babe. Someone to care for, someone to try to please, someone to love. We could all go to the manger at Christmas and be reminded that we have someone special to love, someone divine to care for. Those were our thoughts as we handled what seemed like a thousand details necessary to present a live nativity scene for the community.

We used the wooden fence which had been blown down in the hurricane to build a stable which today is still standing after twenty-three years. The stable was situated near the chapel on the outside of our front fence. The sisters could see it from the chapel. The Bazaldua, Burgess, and Swetish families were faithful participants and organized the schedule so that the changing of costumes went smoothly during each half-hour switch. We had gorgeous costumes in duplicate so that the nativity scene was constant for about three hours each evening. Our donkey, chickens, cows, roosters, and a borrowed sheep made the scene realistic. When weather permitted, we used a real baby in the manger.

For the most part there was no dramatization. The participants just stood quietly as cars from all over slowly came and went during the three-hour period. Many groups would arrive and carol or sing hymns. Some would play guitars. Some years there were flutes and clarinets as accompaniment. One year we even moved the organ out where it could be used. Some years there would be the Christmas story narrated and sometimes the rosary would be said. For the most part the nativity was intended to be a time for quiet, peaceful meditation. I was extremely edified by many of the participants who stood out during very, very cold and windy weather.

There was always something special for the Cuban sisters in the chapel. Perhaps the Skrobarczyk family would come and perform a posada in which St. Joseph and the Blessed Mother go from door to door knocking and asking for admittance. This was quite a touching little ceremony, and the sisters enjoyed it immensely.

We also set up a very elaborate Christmas crib in the chapel. Midnight mass on Christmas Eve was the climax of the Advent preparation for the birth of Christ. Father Claude Valentine, our chaplain, always gave such a beautiful sermon. Always after midnight mass I played the organ while the sisters sang in Spanish, "Gloria al Nino de Belen. . . ."

During that hymn, which they always sang so joyously, each sister approached the altar where Father held the Infant which we had in the crib beside the altar. Each sister went up and kissed the Infant and then after all had come forward Father placed the Infant back in the crib. This is an ancient Cuban custom which we still carry out to this very day. After midnight mass it was into the refectory for hot chocolate, always too strong and too sweet for me, then to bed to rest for the mass or masses and prayers of Christmas morning.

At midday we had our big Christmas dinner . . . turkey with stuffing, cranberry sauce, mashed potatoes, sweet potatoes, peas or beans, salad, some kind of pie, and always a little glass of wine. Then maybe a siesta and vespers late in the afternoon. Turkey sandwiches for supper were followed by the big time of sharing our little gifts. Of course, we did not exchange gifts but the superior always asked the sisters for a list. Then I or Sister Rosa would go out and try to find the items on the list. Each gift was put in a fancy little sack with the sister's name on it.

One year we had a special attraction at gift-giving time. We always had a huge Christmas tree in the refectory, and this year Madre Clara was dressing up as Santa Claus. Madre Clara is very, very short—the result, she claims, of being the nineteenth child in her family—so it was a job to tuck pillows into all the right places. But we finally got her put together, ready to go out and entertain the sisters.

As a surprise we had bought a three-wheel bicycle for the sisters to use to ride around the monastery grounds, and at the last minute Madre Clara said, "Oh, I want to ride in on that bicycle in my Santa Claus suit." So we rigged everything up so that when I started playing "Jingle Bells" on the piano Madre Clara would ride in on this bicycle singing "ho . . . ho . . . ho . . ." and ringing a bell. At the proper moment I struck out with "Jingle Bells," and I heard this bell ringing in the distance and "ho . . . ho . . . ho" getting louder and louder.

Finally in came Madre Clara on the bicycle. Everybody was so excited that you could feel the electricity in the air. I turned around at the piano just in time to see Madre Clara going "ho . . . ho . . . ho" and not being able to brake the bike. It turns out that she didn't know a thing about brakes! The bicycle ran kerplop into the metal folding table upon which rested our gorgeous Christmas cake. The legs of the table on the far side of Madre Clara folded under ever so neatly and the cake came sliding down that table and slid right across the floor. The three kings riding their camels on top of the cake never even wobbled. Everybody thought Madre Clara had planned that, and she and I never put them any wiser.

One of the most enjoyable moments I ever experienced at the Corpus Christi monastery occurred the day Abraham Gonzalez, our teacher from the music store, came out to conduct Sister Angela and me in our first clarinet concert. After we had been given the instruments for Christmas, I said to Sister Angela, "We need to take lessons so that we can be playing by the time Sister Celina has her seventy-fourth birthday. We can give a concert as part of the celebration!" And today was Sister Celina's birthday. Abraham was coming out to conduct us in our first concert.

"Have I ever got a surprise for him," I said, pointing to our teacher as I whispered conspiratorially to one of the monastery's friends who was there for the special occasion. "I bought a saxophone at a junk store a couple of weeks ago and have been practicing. No one knows that I have got it."

So after Sister Angela and I performed the expected numbers on the clarinet and the keyboard, I hurried out of the room to get the saxophone. Returning, I tore into a pleasant enough version of "Maria Elena," I suppose, given that I had had no instruction and time for only four practices. "Now I want you to teach me "Am I Blue," I told Abraham Gonzalez. "That'll sound good on the sax." Then I began to pantomime and sing, "Am I bluuuuue . . . ," and all the nuns had a good laugh.

"To close this event," Abraham spoke in Spanish immediately following my silly escapade, "I want everybody to get involved in this concert. I want every nun in the monastery to perform."

The Cuban sisters sat there shyly, not even looking up at the speaker. They knew they weren't going to perform, no matter what this music master suggested.

Then came the surprise. While Sister Angela and I played one song on the clarinet, Abraham played another. Our song was "America the Beautiful." Abraham's, the Cuban national anthem. Suddenly, every American in the room was standing, singing along with Sister Angela and me. And every Cuban nun in the house was on her feet, singing to the top of her lungs, tears streaming down her cheeks, to Abraham's accompaniment. To hear their national anthem so unexpectedly had moved the Cuban sisters deeply and brought them immediately to their feet. And I must say the simultaneous sounds of the American and Cuban anthems were beautiful; it was amazing how well the two songs blended. The experience bonded us all more than ever. For anyone listening could very well distinguish the Cuban words and melody and at the same time hear "America the Beautiful." Yet the harmony was just amazing.

CHAPTER SIXTEEN

GRACES OF INTERIOR PRAYER

My heart overflows with noble words.
To the king I must speak the song I have made;
my tongue as nimble as the pen of a scribe.
FROM PSALM 45

I will praise you, Lord my God, with all my heart
and glorify your name forever;
for your love to me has been great.
FROM PSALM 86

I don't remember exactly when I knew that I wanted to belong to God alone. Perhaps it was kneeling in front of the altar in

Mama's room as a little girl. Maybe it was making visits to the Blessed Sacrament at noontime while I attended St. Peter's School. Could it have been that night in high school when I went with Pete Starn to my first formal dance, when I realized that it was more important for me to fast after midnight in preparation for Mass the next morning than to go with Pete to get something to eat after the festivities? Maybe it was sitting up in an apple tree before I ever even went to first grade, looking out over the beautiful flowers and trees in our yard and feeling at one with all I was seeing. Who knows? Who is to say?

I do know that even as I played the organ for mass in the choir loft in Niagara Falls, especially when playing for a wedding, I felt my own commitment to God so very, very strongly. There had clearly been something like a second conversion after the crisis in my soul that occurred when I was at Bolivar, the crisis that led me to enter the cloister.

As my faith and devotion to God continued to deepen, it seems that I was blessed by being given the gift of the graces of infused contemplation following some inner struggle. Or at a time during which I needed some special graces in order to know how to proceed. Or when I needed actually to be directed to proceed in some particular way.

First, my vocation to the Poor Clares had been preceded by the graces I received in the chapel at Allegany when Father Juvenal was giving me spiritual direction during summer school at St. Bonaventure University. The graces then had lasted perhaps a year or two and then had faded away.

Then, while in New Orleans, they became prevalent again right before the Cuban sisters arrived and continued up through that final experience in the chapel which led to my coming to Corpus Christi. Then the graces had faded again.

For more than ten years now, as we had struggled to get the foundation established, I had been very, very busy. By necessity I had to be at the service of everyone and everything at all times. Consequently, my contemplative life had been held at bay. But recently I had begun to yearn for quiet times. Unfortunately, I could see no way that I would ever be able to have them.

Then came the most unexpected news. Monsignor Thompson came to the monastery from the chancery office, bringing with him a priest who had for some time been a hermit living near Austin. "Father Burke has been assigned as your chaplain," the Monsignor told us. "He will be available to celebrate mass every morning, hear confessions, and provide spiritual direction." It was through this turn of events that I was to find the encouragement I so desperately needed in my spiritual life.

In the months ahead, I once again began to experience graces of infused contemplation. Holy Mass became alive for me. Almost every day, different words in the prayers of the mass took on deep meaning and brought great joy. Frequently, at Holy Communion time, I would be lost to my surroundings and completely absorbed in God. I began to find more time for prayer which seemed to renew my physical strength and sharpen my mental clarity—both of which I greatly needed to face the difficult situations which were quite plentiful during this period. There were times that I spent the greater part of the night in prayer even though I had retired with the intention of sleeping. Every time, without fail, I would be more refreshed than if I had slept soundly.

I always kept pencil and paper on hand because, at times, the periods of infused contemplation would be followed by a time of deep recollection in which I felt an immediate urge to write. The pieces—I came to think of them as poems—which I would put on paper were all written rapidly, and never a word would be changed. Each piece seemed to me to express some aspect of infused contemplation, and, even today, have intense meaning for me. They serve as reminders of God's love and His continued Presence in my soul. When I meditate upon them, the writings keep my spiritual life focused, and I am reminded to "measure everything with the yardstick of eternity."

When returning to myself after one of these prayer experiences, I felt the great weight of measured time, for the timelessness experienced during infused contemplation is a foretaste of life in eternity where, with God, time does not exist.

Ode to a Clock

O face of numbers
And hands that go 'round and 'round
Would that I could throw you
To the winds that blow
And loose your measured marks
That mar my coming and my going.

Today my constant acquiescence to your wishes
Has worn me thin
While in my hunger unfinished meals
Become remembrance not to be remembered
As hunger of your counting
Measures not in measureless ways I know.

Your song is not my song
For your limit sets a stop
Where no end begins the song
That I have learned to sing.
My voice, in silence drowns
The ticking shouts you whisper.

Yes, I run to your rhythm now
But when the ticking of my heart-clock ceases
Your song shall fade away
Having beat me down my path,
Where my running turns to rest
And my time to love knows no limit.

The moments of contemplation were so desirable that only a spirit of poverty could reconcile me to its absence. While I felt renewed strength to continue living for God, at the same time, I knew I did not deserve nor did I have the right to claim the Presence of God in this manner. I could only hope and wait for these precious moments of contemplation:

Emptiness

To hold myself in readiness for Your coming
Even though I know that You are here.
This, the only right remaining here within
The waiting vessel held for feast
Or famine, no right for me to say.

My work to keep the empty
Open vessel ever empty
For its Maker only
Has the right to fill
And my longing thirst
Can never claim the water
By filling it with tears
The earthly senses cry,
But hold it still for feast or famine
For only He in love can say.

Oh, I lie and scan the paths within me
For His coming
But as I wait, of sudden He is here
Never walking down the ways
And trodden paths of yesterday
Nor casting shadows
On His lawn of tomorrow
As He comes in love
Of feast or famine.

During the state of infused contemplation sensate knowledge—knowledge from the five senses—may be paralleled with an experience of one or more of the five spiritual senses:

Conversation Between the Sense and the Spirit

1

"How can one speak of Love
Without singing of the Lover?
Lips speak, hearts sing!"

"But I
Dumb-struck; numb-plucked."

2

"How can one see the Light
Without knowing the Sun?
Eyes see, minds know!"

"But I
Blind-glow; dark-flow."

3

"How can one hold the Love
And not feel the Lover?
Arms embrace; hearts throb!"

"But I
Sense-nipped; spirit-dipped."

4

"How can one detect the Perfume
Without smelling the Rose?
Honey dew. Liquid delight."

"But I
Permeated; saturated."

5

"How can one relish the savor
Without tasting the Food?
White Bread; red Wine."

"But I
God-dined; Love-find."

And, of course, God cannot fill a heart that is already full of things not Himself. Our hearts must be kept free by a sense and a spirit of poverty. There must even be a detachment from contemplation itself because these graces are never an end unto themselves.

Poverty

I know it is no longer there
that to which I held with all my might.
And so in endless graspings
I have fallen flat upon myself
and lying here gaze up into the blue
through empty air and wait, and wait, and wait.

Oh yes, I see my empty hands before me
but I wonder what they could have held
that cost me so.
Now I lie and wait for other Hands
to lift the pile of nothing I call "me."
And where they take me now I care not.

For I have been to all the places
man could be.
Now I wait for I have heard a promise
that eyes, and ears and mind have not yet known.
I lie and wait, for eyes and ears now serve not
to grasp the unknown Maker of my mind.

When Father Michael was transferred, that feeling of interior stripping that is part of the poverty of the soul greatly intensified:

That I May Leave You

Now in peace I lie upon the pile of nothings
And let slip through the fingers of my mind
The things that were or could have been
The "now" is here. I find it good

To put you where you are at last
No more to play with things of time
That I may leave you in His Heart
For tonight I shall die.

I breathed your air a hundred times
And kept my engines going
With eyes upon the road ahead.
And hands upon the wheel of time
I drove my bargain toward His goal
But ran aground upon His sand
That I may leave you in His Heart
For tonight I shall die.

In silence now I cannot hear
The silent shouts of things I see.
Unspoken words ring loud and clear.
My mind's eye sees them there
Upon the surface of no plans
To clutter up my journey now
That I may leave you in His Heart
For tonight I shall die.

The wheat and wine have now become
The only Food I need
Since hunger lost is hunger gained
While thirst for Him is drink.
So offer not your earthly food
No need have I to eat
That I may leave you in His Heart
For tonight I shall die.

Boundless joy, delights untold
Await me now I know
Upon my painless bed of pain
As I arise and go. The place
You know, you told me so
That I may leave you in His Heart
For tonight I shall die.

As another unexpected turn of events would have it, I became
acquainted with Father James J. McQuade who was to have a
profound impact on my life through his spiritual direction. A
dear friend of mine from another monastery was engaged in a
directed retreat with Father McQuade, and even though I had
never met Father, I decided to write to him to tell him that I was

praying for Sister. On the spur of the moment, I enclosed these
untitled words which I had written:

> Eyes wide open, seeing nothing
> I cling to it in blindness
> And in anguish yearn for His darkness
> Where Living Water runs deep
> Knowing well that He awaits my plunge,
> All thirsting for that quenching moment
> I dive into the pool of anguish
> And trembling sink into the nothingness
> To wait, wait, wait unknowing
> And unremembering how to drown.
> I find that I have died but live
> To Him alone Who hides my Life.
> His Blood wet around me
> There is no other place to be
> His Body held in human hands
> Faith screams at me and knocks
> Me down to lift me up
> Where all things rush in place
> And resting now with arms
> Outstretched so wide apart
> That no embrace of mine
> Casts shadow on my heart for
> In faith laid bare my nothingness
> All mind-stripped, will-pierced I hang.
> Beaten, shattered, split and crushed
> There's nothing left of me
> So go Your Way and take Your time
> Whatever be Your Will.
> My time is short
> His has begun
> I've lost the battle
> My God has won.

Father sent me a short note, thanking me for my prayers for
Sister. Inside the envelope was also tucked a copy of the lines
that I had sent him, but with one change. Father McQuade had
provided a title: "Contemplation." There are no words for what
his adding that title meant to me. This priest, who was a
stranger, understood. He knew what was going on in my soul.
With that one word, "Contemplation," Father had confirmed my
experience. Later, when I went to Michigan to make a retreat

under his direction, he asked if he could hand out copies of my lines to his retreatants.

It was on one of the several thirty-day retreats I was privileged to make with Father McQuade that I experienced an inexplicable moment. Words capture the experience so weakly. I had just finished my confession and Father raised his hand to give me absolution. Just as he began, something took place. It happened and lasted for only an instant. I could not have stood it any longer. I was allowed for a piercing instant to see the *beauty of my soul*. It was a moment of piercing delight, of reverent awe, of astonished reality, and a great fear of harming something so beautiful and so fragile as that which I carried about within.

The experience brought about an immediate reaction within me, and I gasped. Father hesitated for a moment and quickly looked at me. I said, "Oh, Father." There was just a moment of silence and Father very casually continued the absolution. It was as if he understood exactly what had happened. Later when I went over to sit down on the other side of his desk for our hour of study, I said, "Father, something happened then, didn't it?" He answered, simply, "Oh, yes," and we went on with our study. Since that momentary experience, I have tried to compare the beauty of my soul to fragile and beautiful things I know and can see. Later I prepared a slide presentation which was suggested by this experience of the inner vision of the *beauty of my soul*.

The memory of these experiences of contemplation give me consolation when "the well runs dry" and I must plug along in pure faith without being aware of God's Presence—which is most of the time. The periods of infused contemplation last a year or two, and then for ten to fifteen years He is silent within my soul. Actually it is in the dry times that one has a chance to prove one's love for Him. When all is going well and God's gift of experiencing His Presence makes all things easy, that is one thing. It is also by His grace that perseverance is possible while walking the arid desert in dryness.

Strengthened by His graces, one is able to carry on during the most difficult trials. I had no suspicions that the coming month would hold such excruciating moments and that it could be only His graces which would sustain me during those painful days.

It was December 1976, and Sister Rosa and I were taking the monastery's Christmas gifts around to all the benefactors. Over the years my love and friendship with Sister Rosa had deepened. First, of course, there had been that strange experience of total revulsion which I felt for her when the Cuban nuns first came to New Orleans. This revulsion, however, had been transferred into the sweetest and purest of love in that single moment on the stairs when I chose Sister Rosa to be my partner in the art room. From that moment forward, Sister and I had experienced an unbreakable bond which often was a source of solace as we struggled to establish the Cuban sisters in a new foundation. Only for the time that I was in New Orleans after Sister Rosa first went to Corpus Christi did we not kneel every night as we had that first time in the art room in the monastery in New Orleans to say our Hail Marys together.

Around Corpus Christi we were a familiar sight—I driving the monastery car and Sister Rosa jumping out at every stop and running a gift into the doctor's office or into the hospital or to the lawyer's or the Bishop's. I sat at the wheel, waiting.

When we reached home, Sister Rosa said she didn't feel well. "You're over tired," I told her. "Get a good night's sleep so that tomorrow you will feel rested." During the night, however—about two o'clock—one of the sisters came to my room saying, "Sister Rosa is ill." I hurried to her room and knew within seconds of looking at her that she was having a heart attack. "Call the doctor and an ambulance," I instructed someone, almost in a panic. "I'll follow to the hospital in the car with Sister Holy Spirit."

The heart attack was quite severe. Sister Rosa remained in intensive care. For several weeks I lived at the hospital, running home as fast as I could in the car to take just enough time to have a bath, change my clothes, get a bite to eat, and hurry back. Every day I said the rosary with her. But because she needed so much quiet, I would say only the first half of the prayer and then we would just keep silent while in our hearts we answered each Hail Mary and Our Father. I read to her very little because she needed and wanted to be quiet, so most of the time we just sat together, she in her bed, and I in the chair right beside her. Once she asked me to bring my autoharp to the hospital, so every day after that, in the afternoon when the rosary and our quiet meditation were completed, I would play the autoharp and sing softly whatever hymns she requested. The last hymn I recall singing for her was "Breathe on Me, Breath of God."

Sister had a great devotion to the Blessed Mother. Of course, we continued to say the three Hail Marys together every day, but once Sister Rosa asked me when we had finished, "When you know that I am dying, will you please start talking about the Blessed Mother?" I promised her that I would do this.

It was now approaching late January and Sister had been in the hospital for more than a month. On this particular morning I hurried in after my quick trip home to change clothes, and Sister said, "Oh, I missed you so much while you were gone. Please don't leave me today. Stay in the room."

At noontime, Sister Rosa said, "I don't feel like eating this hamburger. You please eat it and then you won't have to go down to the cafeteria." After the nurse took the tray away and Sister and I said the rosary, Sister Rosa said suddenly, "Call the nurse." I called her but she did not come right away, so I went to the door to see where she was. When I looked out, instead of the nurse, I saw Father Haas, one of the Schoenstatt Fathers who was known for his great devotion to the Blessed Mother. Appearing at the door, he said, "Sister, I got off on the wrong floor by mistake and then I remembered that Sister Rosa was on the tenth floor, so I thought I would just run in to see her."

I replied, "Oh, Father, that's fine. Do come right in. Sister isn't feeling well." When he reached the side of the bed, Sister Rosa said, "Father, please start talking about the Blessed Mother." After the brief conversation in which Father Haas honored Sister's request, he left and shortly thereafter she put up three fingers. I knew that meant she wanted me to say the three Hail Marys. As I finished the last one, Sister closed her eyes and her head slumped to the side. I gathered in a tissue two tears from the corners of her eyes and this tissue I kept for a long, long time.

At Sister Rosa's funeral, I knelt at the coffin for perhaps fifteen or twenty minutes before the Holy Mass, taking a rosary or a holy picture or whatever a mourner handed me, placed the object in Sister Rosa's hands, said a Hail Mary, and handed the item back to the individual. I had asked the priest if after the Holy Mass, I could say a word or two—I felt as if I just had to be some part of that ceremony—and I recited a poem, "The House with Nobody in It." It struck me so forcibly to see Sister Rosa's body there in front of me. During her lifetime that body had been the resting place and the temple of the Holy Spirit; now it was empty and lifeless.

One day, I was missing her very much and just wanting to talk to her. "But it seems like such a one-sided conversation," I was thinking to myself, when suddenly I heard her say, "Measure everything with the yardstick of eternity." I don't know where those words came from, but they impressed themselves indelibly on my mind and my heart. So many times since that moment, when things are not going well or when a particularly hard task is before us, I am buoyed up by those words: "Measure everything with the yardstick of eternity."

I frequently remember now the conversation we had a few days before Sister Rosa's death. "Promise me something," she had said suddenly one afternoon. "Certainly, Sister, if I can do it," I had responded. "Promise me you will never abandon my sisters," was her request. "I never will," I answered. Even now I know that Sister Rosa has not abandoned this community, nor ever would I.

CHAPTER SEVENTEEN

OF COURSE, HORSES! THIS IS TEXAS

How many, O Lord my God
are the wonders and designs
that you have worked for us;
you have no equal.
Should I proclaim and speak of them
they are more than I can tell!
FROM PSALM 40

It was a beautiful day. "Great weather to do some horseback riding," I told one of the postulants. "Let's saddle up the quarter horses and ride along the back of the land."

"This is the snake path, you know," I warned the postulant as we jostled along. "There's one big snake in here that has managed to continue to outsmart us. So be careful."

About that time, she yelled, "And there he is!" She was already on the other side of the snake and I was approaching. I couldn't see a thing but suddenly Macetta, the big Appaloosa mare I was riding, did something I had never seen a horse do. She bent down in front and stretched out as straight as she possibly could. It was all I could do to stay on her. Then suddenly she sprang off to the right and I did, too. I had the sensation that I was just shooting off into the air; I found myself looking up at the pretty blue sky and the clouds and then feeling the hard ground underneath me as I fell flat on my back, the air knocked out of me. My sneakers had come off—thank goodness, I didn't have on boots, which were my usual attire—and one had landed about ten feet to the right and the other ten feet in the other direction.

The big question, of course, was where was Mr. Snake in all this confusion. Fortunately, as I scrambled up, I saw no snake. But I saw no horse either. I retrieved my shoes and walked slowly up to the house where the horse was waiting for me. Remembering that what you were supposed to do was to get right back on a horse that throws you, I climbed up on her back and rode around a bit in the yard. But already I was getting sore from the fall. And reality, in general, was sinking in. Riding horseback could be a dangerous thing and maybe, at my age and with the responsibilities I had, just a little foolish. Although I loved riding so very much and did a lot of it, I recognized that I could be seriously hurt. So not too long after the snake scare and the fall, I decided to sell the horses.

But, oh, how the decision hurt. Every day I would look out across the back of the monastery property to spot the horses; then I would realize they were no longer there. Horses had been a part of our family as long as I remember. Why, even on my grandfather's tombstone in New York were engraved the words, "Adolphus Muller, Father of Carriage Designers in America." This grandfather had come from Stuttgart, Germany, spoke eight languages including Japanese, as a draftsman designed horse carriages of every style, published a bimonthly art journal for carriage and wagon makers, and was a masterful artist. My father took after him.

Then my great-great-grandfather on my mother's side was Major General Winfield Scott of the United States Army. His image on a horse I always had hanging in front of me on the wall of the monastery office. A lovely Duvall lithograph from life on stone by A. Koellner showing Grandfather Scott mounted on a flashy dapple gray stallion.

My maternal grandmother, so they tell me, rode sidesaddle over fields, fences, and through streams. My mother drove a horse and buggy to school and church. Both my sister Mary and I rode horseback. And some of my earliest memories are of the great times we had when my sister would drive my brother and me back to Bargaintown Lake where she would take the pony, still hitched to the cart, right into the water. She would tie him to the overhanging tree limb and then, wetting a sponge, put it into a straw hat which she plunked on the pony's head. (It stayed there because she had cut holes for his ears.) This kept the pony cool while we had a swim. We would then drive home through a wooded path flanked by pink and white mountain laurel in bloom. "Ah, those were wonderful days," I would think as I stood looking out at the now-empty pasture, the family memories making even more poignant the now total absence of horses from my life in Corpus Christi.

"Don't you miss the horses?" Madre Clara often asked me. I tried hard not to let on. "It wouldn't do anybody any good to know how *much* I miss the animals," I reasoned with myself. "This is just something I will have to get over."

But then one day someone dropped a horse magazine off at the monastery and, as I was looking through it, I saw for the first time ever a picture of a miniature horse. "Ah, our pasture looks so empty," I said to myself. "One of these tiny horses would make such a darling pet." Showing Madre Clara the picture, I said, "I think I'll phone just to see how much these little things cost."

There were three ads in the magazine. "Sir, do you have some miniature horses for sale now?" I asked the first voice that answered. "Sure do!" came the southern drawl. "Could you tell me the price?" "Three thousand up," came the answer. "Uh," I stumbled in response, "we just wanted a pet." "That's the starting price," the voice answered. "Three thousand dollars."

I went on to the next ad.

"Sir," I asked, "how much are your miniature horses?"

"Three thousand up," came the answer.

"Sir, would you like to donate one to a group of Poor Clare sisters?"

"No, thank you, ma'am," was the reply. So that was that.

I dialed the number of the last ad. This time a woman answered the phone. I told her who I was and then lost no time beating around the bush. "Could you donate a little horse to the sisters, ma'am?" I asked her. This is what I heard over the phone: "Sister, I am not a Catholic; in fact, I am a Baptist. But I would love to donate a little horse to the Poor Clare sisters."

"Jackpot," I yelled out to Madre Clara. "I love the Baptists!"

"You will have to pay for the crate and transportation, though," the woman told me, and since we did not have the money even for that, I called Jon Held, the contractor who had built the monastery.

"Yes, I'll pay the freight," he said, "and I'll buy you a mare also. If you are half as good at raising horses as you were at raising birds and cats, you've got it made," exclaimed Jon. And that was how we got into the miniature horse business.

We borrowed Tom Ryan's truck and Johnny Garza drove me to San Antonio where TWA unloaded two small crates onto the pickup. We could barely get a glimpse through the slats. What excitement when we arrived back in Corpus Christi. The little chestnut stallion stepped out of the crate straight into our hearts. The little pinto mare was precious. Countdown and Ginger, we named them—our first pair of miniature horses.

The little tag on my tea bag the next morning at breakfast read, "The future is purchased by the present." Of course, I had no way of knowing then how true that statement would turn out to be, for it was to be the miniature horses that would allow us to take care of ourselves when raising birds and cats was no longer possible.

On the way home from a retreat with Father McQuade in Michigan, I decided to change my plane ticket and get off in Columbus, Ohio, to visit the farm that gave us our first mini. It was there at Bob and Fredericka Wagner's farm, as gorgeous little minis were paraded before my eyes, that I realized the potential of the miniature horse.

"Group those fifteen horses there in lots of five," I told Freddie. "Depending upon how much money I can raise, we will buy one, two, or three of the groupings."

Once again the Catholic Women's Fraternal of Texas came through. With a loan from them we were able to buy the horses and build a lovely little barn complete with a small office and living quarters. Johnny Garza gave up his job as a mechanic to come train the horses.

Channel 3 television heard of the pending arrival of the monastery miniature horses. They were there with camera rolling when the enormous van pulled in. I had never seen such a horse trailer. It held our fifteen minis as well as many of the Wagner's horses which had just been shown at the fair in Dallas. The film of the arrival of the horses caused quite a stir. When we had an open house (or perhaps I should say an open barn), a huge crowd showed up for the event. Father Valentine, our chaplain, blessed the barn; my sister Mary gave information on the horses as I introduced each to the public while Johnny Garza handled them. We now had sixteen mares and two stallions, well on our way with an immediate breeding program.

One day I received a call from a gentleman in Dallas who told me he had purchased a miniature horse for his wife for Christmas. "I think it's a nice looking horse," he said. "Would you be willing to show it for us?" That began a long association with Bob and Sandy Erwin who, after their horse Egyptian King shot to stardom the first time we showed him, started buying small mares which we kept in our stables. Soon they owned forty minis. With Bob's financial help, we built a beautiful sixty-stall show barn and Joe Cordova added a huge arena. Bleachers were in two niches in the front of the barn with an entrance into the arena from the barn between the bleachers. Over the entrance was the announcer's box. We sponsored three American Miniature Horse Association shows during Buccaneer Days in Corpus Christi. Several thousand people attended.

With all the Erwins' horses and ours, the twenty acres of monastery seemed almost weekly to be shrinking. Perhaps this was why I received the news of the possible loss of our Flour Bluff property as calmly as I did, although I did everything I could to keep it from Madre Clara and the other sisters.

Visiting our new land with Madre Clara, Christine Knapek, Sister Bernadette, Benita Pavlu (President of the Catholic Women's Fraternal of Texas), Sister Rosa

I prepare to cut the grass on the new monastery grounds— if Paul Swetish ever gets that old mower fixed.

We clear land for the new monastery in Corpus Christi. Madre Clara shared a joke on herself. She had thought Tom Graham's bull was going to consume all the underbrush. "Bulldozer" became a new word in her vocabulary the day Tom came with his machine instead of a bull to clear the land.

Watching the finishing of this cement slab in Corpus Christi, little did the sisters know there would be still another monastery to build!

The Poor Clare monastery in Corpus Christi, completed May 1966.

We were a Poor Clare rock-and-roll group and made our own good times. We thought we were great! But, then, we had no audience but ourselves!

Fun in Corpus Christi—atop one of our four riding horses.

My 410 shot gun, used to kill rattlesnakes. The trap set for rabbits caught a skunk instead, so that night I waited for the rabbits in the dark and shot two in one shot by sighting only their white tails!

It was my chore to milk the three Nubean goats twice a day. Later, we had cows.

A sense of accomplishment gave Madre Clara much joy as she gathered fresh eggs from our own hens and milked our own cows.

Sister Rosa and I pick peaches in Corpus Christi. Oh, that cobbler was good.

Here I am riding the famous white donkey, Cola, mother of Coke.

Sister Caridad finally received the crown of thorns and took her vows before
Madre Clara at age eighty-three. What a day of joy for all of us.

This is one of our Himalayan blue point cats. We raised Persians also. We had 75 queens and 12 toms and shipped to pet stores all over the United States.

The sisters in Corpus Christi.

Front Row, left to right: Sisters Caridad, Magdalena, Rosa, Margaret Mary, Mary Bernadette, Clara, Espiritu Santo, Encarnacion, Corona.
Back Row, left to right: Sisters Celina, Ascencion, Luisa, Mary Theresa Clare, Veronica, Martha, Claudia, Patrocinio, Teresita, Rafael.

Taken about 1970, this photo shows not only the Corpus Christi community but also the statuary representing our Lady, Patroness of Cuba. This heavy statue, measuring approximately 36" x 36" x 32", was carried on the cattle ferry to Miami by the Cuban sisters when they evacuated Havana in 1960. It now has a prominent place in our refectory in Brenham.

BRENHAM, TEXAS
1985-present

Cowboy Nun From Texas—Home At Last!

CHAPTER EIGHTEEN

CONFLICT AND JOY

The clouds poured down rain,
the skies sent forth their voice;
Your arrows flashed to and fro.
Your thunder rolled round the sky,
Your flashes lighted up the world.

FROM PSALM 77

Waldron Field, a "touch-and-go" practice field for Navy pilots, had been reactivated since we had built the monastery on Yorktown Boulevard in Corpus Christi. Our monastery was located at the end of the runways, just across the street. "This is a declared danger zone," the Navy informed us after the field had been put in use again. Many articles appeared in the local paper concerning properties in the area. Clearly, something was happening.

"What in the world are we going to do?" Madre Clara kept asking me. "I don't know," I would tell her, "but if we do have to move, I know God will provide for us." We lived day to day, not knowing what our fate would be. That within itself was extremely unnerving.

Our friend Charles Kaler, efficient helper that he was over the years, once again came to the rescue. "We'll bring things to a head," he said. "I'll badger the Navy until we reach the proper authority."

The Navy turned out to be wonderful. Charles was able to make clear to the authorities how untenable the situation was for a group of contemplative nuns. "They desperately need to know what action is being planned for this section of the community," he repeated to the appropriate authorities. Following these conversations, transactions for the Navy to buy the monastery and land were carried out quite smoothly. Two very kind women in charge of Navy real estate transactions came to help us plan the actual moving of the entire operation.

But first we had to find a new location.

We could not find satisfactory land near Corpus Christi. We did find a ranch for sale in Luling, but after Bishop Harris had lovingly accepted us into the Austin Diocese, we discovered that the property was just outside the diocesan boundary line. Bishop

Rene Gracida, new that year to the diocese of Corpus Christi, suggested that we find land in the Austin Diocese since Bishop Harris had been so happy to have contemplative nuns. We began a serious search then for land in the Austin Diocese.

Of course, our major concern was the spiritual and physical welfare of the sisters. We had to have a contemplative atmosphere. That was a must. Then we had our livelihood to think about. We had to have room to raise the miniature horses.

"Look at that hill," Sister Mary Joseph exclaimed excitedly as we approached property number seven in our search for the right location. "Can't you see the monastery there?" But then we came to a second hill, better still. "Look," I cried. "There." Surrounded by a wooded area was this hill, waiting for all eternity for our monastery. It was at that moment that I remembered hearing those words in the chapel in New Orleans: "Why are you afraid to come to me (in Corpus Christi)?" The meaning now of those parentheses I had somehow heard became clear. Moving to Brenham had already been sanctioned by the Almighty. Suddenly I knew it would not be hard to leave everything behind— the property, the building, our friends. We would miss them, sure, but I knew now without a doubt that this move was certainly God's Holy Will.

Sister Mary Joseph and I returned to Corpus Christi to get Sister Angela and Madre Clara. They were every bit as excited as were Sister Mary Joseph and I. But there was one hitch. Twenty-five thousand dollars earnest money was required right on the spot. The Navy had not yet paid us, and we had no reserves in savings.

"Let's go back to the motel and pray about this," someone suggested. So back to the motel we went. I went to my room and made a phone call. "You know those two best mares you have been trying to get me to sell you," I said to the man on the other end of the line. "Now I am ready." "How much do you want?" he asked. "Twenty-five thousand dollars," I replied. "You'll have it in the morning mail," the voice answered, without one hesitation. "Where would you like me to send it?"

I stepped out of my room and reported, "We have our $25,000," I told them, a big smile spreading all over my face. "Let's go back out and look at our land."

The $25,000 down payment was made and the contract signed, with the remainder due December 31, 1984. This was early June. The Navy was expected to buy the Corpus Christi property by early fall so we would have the money for the land in plenty of time for the closing. "I think I'll ask the owner if we can go ahead and construct a guest house on the property," I checked with Madre Clara. "That way, we will have a place ready when it is time to oversee the construction of the monastery." Brenham, in Washington County, was to become our new home.

The owner agreed to our request. We made arrangements in the late fall for a small trailer belonging to Don and Flo Mertens to be pulled onto the land and connected to an electric pole and an existing septic system. Sister Mary Joseph and I moved into the trailer to watch the guest house go up.

The weather was rainy and cold. We kept the trailer heated but there were plenty of drafts. Our beds were side by side, and we piled the covers high. At the foot of the beds was a small shelf, enough room for an alarm clock and a tiny black-and-white television which we sometimes used in the evenings.

One night I was sleeping soundly and having pleasant dreams. Suddenly, there was a terrific blast, and lights were flashing at my feet. I leaped up into the air and hit the ceiling in the tiny trailer. Quickly I turned to look at Sister Mary Joseph. She was not sleeping. She was not lying down. She was not frightened. Instead, she sat there with a scowl on her face, arms folded like a tough school marm. "What happened?" I cried. The television was screaming at me, all loud static and bright flashes. I looked at her, then back at the television. Finally, of course, I turned the thing off.

"You were snoring so loud I could not sleep," she told me.

"Now that makes two of us," said I.

Then we both doubled over in laughter.

I must have behaved from that time forward for that was the last of the midnight television.

Daily, we drove into Brenham to the post office to check for the arrival of the expected payment from the Navy for the Corpus Christi property. Something had caused a hitch in the paperwork, and the check, due in the early fall, still had not arrived and it was mid December. Time was fast running out. We had to have that Navy payment by December 31 because that was the date we were due to pay the owner the remainder owed for the new monastery property.

"Are we going home for Christmas?" Sister Mary Joseph kept asking. "I don't know," was all I could tell her. "I certainly had planned for us to go, but how can we leave here without some assurance that we can hold onto our contract? We could end up losing not only the land but also the new guest house we are building."

One morning I thought of a solution. Off to Johnny Lacina, the lawyer, I went to have him draw up a one-month extension to the contract. We mailed the extension to the owner. Not getting a response, I finally decided to approach him directly.

"I'm not signing that extension," the man exploded. "If you do not come up with the cash by December 31, I'm considering selling to an investment company who will give you first option to buy—if they decide to sell." These words hit me like a sledge-hammer. I jumped in the truck and headed for the bank. "You stay here in the trailer," I told Sister Mary Joseph. "Stay here and pray your head off."

Explaining the owner's unexpected action and noting the pending arrival any day now of the check from the Navy, I asked the banker for a short-term loan and he kindly accommodated our needs. I thought the owner would be pleased when I went to his house to tell him that evening, but I found instead that his attitude had become completely unfriendly. All I could figure out was that he had thought he would be able to retain our $25,000 and become the owner of one very nice new guest house when we could not pay off by the thirty-first of December.

Sister Mary Joseph and I went home for Christmas, and then Madre Clara and I returned to keep our meeting for the settlement of the contract at Stone Abstract on December 31, right on the appointed day. Unexpectedly, two former owners of the land were present since they had liens on the property. A veterinarian also had a judgment. These people were paid off and the owner received what was left. A new contract gave us possession of the land and allowed two months for the owner to remove his horses and his property from the premises.

Madre Clara and I returned to our little trailer but not until we stopped to shop at the local supermarket. This was New Year's Eve and did we ever have something to celebrate. I bought a nice bottle of champagne and Madre Clara bought twenty-four grapes—exactly twenty-four grapes. I stood there completely puzzled as she counted them. Back in the trailer we put our wares in the tiny refrigerator and opened a can of soup for our supper.

We said our prayers and then waited quietly for midnight to celebrate. At 11:30 we gave thanks for all the blessings of the past year, especially for our new land. We prayed God to bless the coming year and the huge construction project that awaited us. Then Madre Clara took out a prayer book in Spanish and proceeded to read aloud. Every once in a while she would stop for me to conclude that prayer with an "amen." I must have said at least a dozen "amens," with one eye on the clock and another on the refrigerator.

On the dot of twelve, we opened the refrigerator door. I took out the champagne and handed Madre Clara the mysterious twenty-four grapes. Then I found out the solution to the grape mystery. "It is a custom," said she, "to eat one grape in thanksgiving for each month of the past year. This is an ancient Cuban custom." It was a new custom on me—I preferred the champagne myself—however, I waited patiently and proceeded to pop one grape at a time into my mouth as she very ceremoniously modeled for me.

Ah, now for the champagne. But it wasn't all that easy. I could not remove the cork. Madre Clara came over to help me. Standing in front of the tiny sink in the tiny trailer, we pushed and pulled and tugged at that stubborn cork. Suddenly the cork began to rise; it was pointed right at the light over the sink. Together we screamed and made for the tiny (and I do mean tiny) bathroom to hold the bottle over the tub. We were both wedged in there holding onto that bottle for all we were worth. Suddenly it shot off with a loud pop, followed by a generous spew of champagne. It scared us half to death but we never let go of that bottle.

We were ready for bed, so we got two enormous glasses, sat on the sides of our beds with the aisle between us. The space was so narrow our knees were touching. Again we thanked God for blessings, touched glasses, and downed the drink. Madre Clara finished before I did and held out her glass saying, "That was good. Pour more. Does it have any alcohol?" "No, not much," said I to Madre Clara. I took another half a glass and Madre Clara finished up the bottle. Needless to say, we both slept soundly.

On New Year's Day we took the contract down to Corpus Christi to show the sisters and then to put it into the safe. I immediately went back up to Brenham. I moved into the newly built guest house to wait for the owner to move his goods off the property so that we could get on with our building. My sister

Mary came to stay with me since all the sisters were needed in Corpus Christi to prepare for the coming move.

One month passed. Two months passed. It was time for the owner to be off the property. The third month arrived and there were absolutely no signs that the horses and equipment and storage items were being moved.

"Is this land really ours?" I kept asking myself. The owner avoided us at all times, but I knew we had to talk. Finally, I suggested a plan to Mary, my sister. "We will go up to his house one evening when the owner gets home from work. When he comes to the door, I will ask him when he thinks we will be able to start construction."

"You are being way too lenient," Mary kept insisting. "You sisters own this land now totally, and it is just by your kind graces that he was given two months to remove his horses and equipment."

But I decided to stick to my plan. We went to the owner's house where I said, "We need to start construction as quickly as possible. When do you think you will have your things cleared out so we can get busy?"

"You told me we could stay until June when school will be out," the owner answered.

"No," I responded. "That is not right. The contract said two months, and that was our only agreement."

Then Mary piped up. "Why, she could not have ever said you could stay till June. The Navy has set a limit upon their remaining on the property in Corpus Christi."

Red in the face, he shook his finger in her face and shouted, "You mind your own damn business. You stay out of this."

"You are speaking to my sister, and I do not appreciate that," I responded. "She is just concerned, that is all." With that, I took Mary's arm and we left the premises.

Charles Keese, the architect helping us plan the monastery, came to our rescue. "We'll send him a letter," Charles said, "telling him that we intend on such and such a day to start pouring a cement slab in the metal building. That we must do this in preparation for moving the ceramic supplies from Corpus Christi to Brenham. We'll give him one week to empty the building or else we will do so."

To be sure that the owner read the letter, I decided to hand deliver it. Johnny Garza, our horse trainer from Corpus Christi, was visiting, and he insisted on going with me. I must admit that his presence gave me courage. All this unpleasantness was foreign procedure for me, but business was business. We had no choice.

The door opened and the owner stepped out. "We need to get started somewhere," I spoke as I handed him the envelope. "This letter explains."

"What right does Mr. Keese have to tell me what to do?" he yelled. "Who the hell is he?"

"He is our contractor and needs to proceed," I responded.

Then, as we stood there on what was actually the nuns' property, I could hardly believe my ears as the man shook his finger in my face and shouted, "You get your g--d--- butt off my front porch."

On the walk back to the guest house, Johnny Garza said, "I have never heard anyone speak to a lady like that, let alone to a nun." That exchange had settled things for me. I was through trying to be considerate of the previous owner. To the lawyer, Bill Spinn, I went. This waiting and uncertainty could go on no longer. We sent a certified letter. It was refused. Finally the sheriff delivered a subpoena. That did the trick. We met at the courthouse privately with our lawyers. To me it was an undesirable situation, but there was no alternative.

Bill Spinn spoke. "You have remained on the sisters' property a month longer than stated on your contract. Have you paid the sisters rent?"

"No."

"Then I think $250 is a fair amount. Can you give her a check now?"

"No. I'll do it Monday."

"You are to remove all your belongings from the metal building and your horses from the pasture so that test borings can proceed and the cement floor can be poured."

Several weeks later there was still no sign of the rent money. Then finally one day Sister Angela called from Corpus Christi to tell me the check had arrived there, with "Donation" written on the bottom. It took another whole month before the man and his horses moved completely out, but with each day I knew we were nearing the end of that particular trial and tribulation.

The test borings were now completed, the metal building preparations were moving smoothly, and construction had begun on the ranch office. Johnny Garza brought our geldings with their carts and harnesses up to Brenham from Corpus Christi. Virginia Brooks, a new friend from the community, brought her horse and buggy over, and she, Johnny, and I drove around inspecting the beautiful ninety-eight acres of monastery land.

After washing down the walls, I painted the inside of the horse barn which had been on the property when we bought it, each stall a different color—ice cream lime, bird egg blue, flamingo pink. When the painting was finished, Johnny brought up all the rest of the miniature horses from Corpus Christi. The Navy even insured the little horses for the move. And as soon as the floor was finished in the metal building—now named The Art Barn by Charles Keese's wife Nelda—Didear Moving Company brought up all the molds, kilns, and supplies so that the ceramic business could get into operation as quickly as possible. We were going to need every penny of income we could get.

In fact, after the horses were moved onto the land and Sister Angela had moved up to help me take care of them, the financial commitment we had taken on to build this new home became more and more real for me. I would find that the euphoria of finding the perfect spot, the certainty that we were in God's Perfect Will would, on occasion, be tempered by a flashing anxiety, "What if something were to happen to the horses? How in the world could we make the payments on the loan we had to take out for construction of the monastery and the chapel?" The payment we had received from the Navy had gone only so far, and we were in debt now for years to come. We had only the revenues from altar breads and miniature horses, plus the little we raised selling the ceramics, as security against that debt.

"We must keep close watch on these pregnant mares," I told Sister Angela, as we fed the little animals every morning. "It is now more important than ever that each foal survives and thrives."

"Yes," Sister Angela replied. "It's very different now, isn't it? In Corpus the monastery was paid for and the horses had only to provide us our daily living. But here . . ." and her voice trailed off.

"Yes, here," I picked up the thought, "they have not only to provide us our daily necessities but they must also pay off our debts." After that conversation we both found ourselves checking each horse even more carefully every morning.

Then one perfect spring morning a baby horse was born on the new land. Sister and I were so excited that we were speechless as we stood there among the bluebonnets and Indian paintbrush to gaze upon the new arrival. Almost everyone has cuddled a new-born kitten or puppy, but it seems even more wonderful to cradle a baby horse in your arms and to bury your face deep down into three inches of virgin fuzz, fuzz too new to know contamination of any sort.

It was a clear spring morning and a perfect day to be born. The vet came and pronounced the little foal in excellent health; and after nursing eagerly, our ball of fur lost no time running circles around his anxious mother. He was so small he could run right under her belly.

"Let me hug him again," I said to Sister Angela and the vet as we started to walk away.

"Isn't he the most perfect creature you ever saw?"

"Yes, he's a little miracle on foot."

"Look at that tiny thing buck."

We were all finding it hard to walk away.

On and on it went most of the afternoon as visitors to the grounds discovered the new foal. Finally, Sister Angela and I did the evening feeding and before we knew it night was upon us.

In the warm weather the horses, including the proud new mother in the pasture, settled down under moonlit sky studded with stars. Exhausted from excitement and the day's work, Sister Angela and I slept soundly, so soundly that we did not hear one sound of the severe storm and deluge of spring rain that came during the night. At 6:00 a.m. we awoke to the fresh smell of newly washed trees and grass. "We and all the earth are invigorated," I said to Sister Angela as I looked from my bed out through the window of the guest house.

"Yes, another beautiful day," Sister Angela responded. "I'm headed out to feed." Remembering suddenly that new ball of fur that arrived yesterday, I jumped jealously out of bed in order to glory in our new possession.

Together we arrived at the feeder in the pasture and started to count heads as usual. The new mother was way out among the bluebonnets yet, standing guard over her sleeping baby.

"That foal is resting in an odd position," Sister Angela noted. For a moment we both looked at one another in disbelief. Without another word, we rushed out into the pasture of wet grass and wild flowers. Coming to an abrupt halt, we saw a sight I shall never forget. Hopelessly looking at the pathetic sight stood the little mare. With legs sprawled out in all directions and his face right down in a big rain puddle, our precious ball of fuzz lay—lifeless and cold.

Frantically, I cleaned the mud from his little nose, gathered the tiny limp body in my arms, and headed straight for the barn. With my precious bundle lying on the dry hay at my feet, I stood gazing down on yesterday's miracle and now asked for another. What had taken place last night? My brain videoed a frantic little miniature horse, followed by a distraught mother, running through the driving rain at the crack of loud thunder and frightening flashes of strange light. Had he become so exhausted that he could no longer run?

My suspicions were confirmed when the vet arrived. "The foal has drowned," he said. "I am so sorry." There was nothing he could do to revive the little victim of a freak storm.

"Would it always be like this?" I asked myself standing there. A pasture full of pregnant mares was an enormous responsibility for anyone, I realized anew, but especially for a group of praying nuns. "Would we be able to make a go of this? Would we be breeding miniature horses this time next year? And if not, what in the world would we do?" These and many other questions began to color my daily thoughts and influence my very dreams.

C H A P T E R   N I N E T E E N

OUR EARLY DAYS AMONG THE BRENHAM BLUEBONNETS

> Be proud of his holy name,
> let the hearts that seek the Lord rejoice.
> Consider the Lord and His strength;
> constantly seek His face.
> Remember the wonders He has done,
> His miracles, the judgements he spoke.
> FROM PSALM 105

But I need not have worried, for as always in the past He who had cared for us continued His wonders to perform. It seemed that I was always led to the right people for guidance and help: the right doctors, lawyers, contractors, bankers, accountants, architects, loan companies, veterinarians, and a host of friends who pitched in and helped with whatever was needed. Many times in those early days in Brenham, people would say, "How can a sixty-six-year-old nun manage so much?" That would always make me smile, for in the back of my mind at all times, even when discouraging and anxious thoughts wormed their way in, were those words I had heard that morning in the chapel in New Orleans: "I am your strength."

The construction of the monastery and chapel was moving along rapidly. I now had a new companion as I oversaw the construction and checked on the little horses each day—a beautiful Dalmatian I had named Apache. And, oh, did Apache wreak havoc at the construction site.

"Git outta here," I heard one of the workers roar.

And out from among the studs of the new building, I saw Apache running—a huge sandwich in his mouth. Following close behind was a carpenter chasing the dog as fast as he could, a big broom waving back and forth in his hand. "I do not know how this dog can think up such mischief," I said, hoping to placate the worker as I chased and called Apache. "He does something new every day."

But as Apache hopped into the front seat of the golf cart I was now using to travel from one work site to another on the grounds, I knew I loved him totally, even if he did daily create some new mischief. "I'll make another sandwich and bring it down to you," I told the worker, as my constant companion and I began our bumpy ride up to the guest house.

At the end of a hard day's work I would say to Apache, "Come on, ole boy, let's you and me go swimming." So back to the guest house we would speed—to the little swimming pool we had built there to provide a place of exercise for any sister who wanted to use it.

As we started in one late afternoon, I grabbed a big beach ball Mary, my sister, had sent us. Apache ran up to the side of the pool, trying to see what had suddenly happened to the bottom half of his mistress. He put his front feet on the ball, only to be immediately surprised by its slipperiness. Plop! Apache fell into the water, coming up a few seconds later, looking very surprised. "Here, ole boy," I said to my companion, "I'll get you back up on the side."

As I put him back up on the solid surface, we both laughed. Yes, I did say we *both* laughed. Well, maybe he was sputtering, but I also knew that when he saw me laughing he gave me his first big grin. From that day forward the dog smiled. Things didn't even have to be very funny, either. If the tail had not been wagging, you might have taken that smile for a snarl. But I knew that when the dog snarled, he growled, and that when he was wagging that tail, he was smiling.

Often as I stood watching the workers framing the windows, building the walls, pouring the floors of the new monastery, I would think of my childhood home....

That house which was never built but which built me. Tales are for telling, but in telling there lies still a secret not oft revealed. And all fourteen rooms in which I grew up did things to me I knew I would never forget.

There was the sanctuary of Mama's bedroom which saw me before I saw myself. That bedroom reserved the right to listen to my first conversations with God, and it heard things no other room ever heard. Each bedroom in the house claimed me for a time. But I belonged in that last one. Bedrooms were for dreams, praying, and being sick in, yet also good for harboring surprise contents and a few pranks. Our bedrooms spent half our lives binding the family together. That was enough.

There was something very special, too, about the staircase that linked our nights and our days. It linked upstairs with the big living room but somehow we did more living in all the other rooms.

If rooms could speak, our library would have the most to tell. The library seemed to claim Papa as No-Nonsense King, for it was off limits to four legs. The fireplace had many functions and in winter the smell of burning logs overpowered the summer smell of books. The books. My link to the outdoors and nature. My winters were always summers as I played with lion cubs, caught butterflies from the pages, was fascinated with the armadillo, and swam among turtles and water lilies.

The library did not hold monopoly on secrets of education. Dining rooms are supposed to be for feasting. Ours performed many more functions. It was a dignified and proper room, always smelled good whether it was roast or roses. Fun and games, study and craft. The walls could bounce with prayer, laughter, music, telephone, or multiplication tables.

The two kitchens were kind to everybody, man and beast alike. Kitchen scenery changed more frequently than in any other room and the signs of the time could never be denied there.

The sun parlor (or conservatory) owned no one. Mama owned it. She shaped it, cultured it, until it reflected her. It was putty in her hands. It was Mama. We loved it and lived in it as we did in her.

The long front porch, our doorway to the outside world, cast a different mood on our gatherings. Maybe through its eyes that outside world broke our family shell enough to let in still another reflection of Himself.

I always heard that our house was one of the first in Northfield. Yet I have the feeling every time I think of it that it was always there. Maybe that is why I feel it was never built but instead built me, for I could not imagine it ever being just a house. It was always home.

"May this new monastery feel, too, like a home for all of us," I prayed as my mind returned to the construction scene on the hillside in Brenham. "May the chapel, the refectory, the bedrooms, the parlor shape us so that we continue to live in Your Holiness and Your Pleasure."

"Sister Bernadette," I heard one day when I picked up the ringing telephone, "I have a bus tour coming through Brenham tomorrow, and the people would love to see your miniature horses. May I bring them onto the grounds?" It was Melani Bayless from the Washington County Chamber of Commerce.

"We are still under construction here," I replied, "but if you want to bring them by, I'll be here to let them see the minis."

That was the beginning. Almost overnight we became a tourist attraction. We began to have so many tour buses stop that I had to solicit the aid of a relative of one of the nuns to help guide one tour while I handled the next one coming in.

The ceramic studio was already set up and we had a few things in the gift shop. People on the tours wanted souvenirs. Then the *Bryan/College Station Eagle, The Houston Post,* and *The Houston Chronicle* appeared on the scene and ran articles on the monastery even before the sisters arrived from Corpus Christi. We had already been featured in a splendid March 1985 article on miniature horses with three photos in *National Geographic. Le Figaro,* the French magazine, had sent a photographer and script writer. Before we knew what was happening, we were getting correspondence from France, England, and Germany.

But answering mail had to wait.

It was now April 1986, and we were busy getting ready for the big move from Corpus Christi to the completed monastery in Brenham. Sister Mary Angela and Sister Mary Joseph organized everything on that end. All packing boxes were labeled with each sister's name. Counting the closets in the sisters' rooms, there were forty closets in the new monastery. Each closet was numbered and things belonging to that closet were boxed in Corpus Christi and marked with the respective number. The move was well organized on both ends and went very smoothly.

Now the Great Exodus began. The Navy representative was there to oversee all moving transactions. The Navy also paid for all transportation and even furnished an ambulance to transport two of the invalid nuns. As the sisters were making last-minute preparations, Sister Martha fell, injuring her head. Another ambulance was sent to transport her to Brenham, with the understanding that she would be taken immediately to the hospital for x-rays in case she had seriously injured herself. (X-rays showed nothing except Sister Martha's fountain pen which for a few minutes the technician was sure was a remnant of some previous operation.) The other sisters traveled in private cars.

Everyone was elated with everything they found ready for them in Brenham. I had even bought all new mattresses. (It had been a hoot to transport all those loads of mattresses in our monastery pickup truck from the Sealy Mattress Company in Brenham.) Each sister found her saint's nameplate attached to the door of her individual room. Even so, some of the oldest nuns still took a while to find their way back and forth to their rooms and around the new, rambling building. "Sister Raphael is lost," someone would call out. "Somebody go find Sister Teresita," another would say. We laughed a lot as we all got used to our new quarters.

We had left behind seven Cuban sisters in the Corpus Christi cemetery, including beloved Sister Rosa. Three sisters had transferred early on from Corpus Christi to a monastery in Mexico. That left ten Cuban sisters now ensconsed in their new home in Brenham.

"What would make the sisters feel most at home?" I asked myself the day they arrived. "What would provide a sense of comforting continuity?" But I need not have worried about this, for the sisters knew how to provide their own continuity. "We put crucifix here," Madre Clara instructed, pointing out a place on the chapel wall behind the Blessed Sacrament. That accomplished, she hurriedly padded down the long hall to the refectory, I close behind her. "Here put Nuestra Senora de la Caridad!" She had found the perfect place for that statue of the Virgin Mary which was well over a hundred years old, the one the sisters had managed to bring on the ferryboat when they left Cuba. "Look good right here in refectory." And I had to admit it did indeed look good in our "family room," the place where we all eat, relax, and rest. By this time all the Cuban sisters had gathered around their beloved statue and begun to praise the Blessed Virgin Mary.

CHAPTER TWENTY

FRANCISCAN JOY

> All of these look to you
> to give them their food in due season.
> You give it, they gather it up:
> you open your hand, they have their fill.
>
> I will sing to the Lord all my life,
> make music to my God while I live.
> May my thoughts be pleasing to him.
> I find my joy in the Lord.
> FROM PSALM 104

Life for me now in my seventy-third year has achieved an almost amazing and a most satisfying round. When I stand on the hillside here at the Brenham monastery, looking out across the rolling countryside, the leafy trees framing the bright red and blue wild flowers and the mini horses in the fields, I could almost be back in that beautiful yard of my childhood home on Jackson Avenue. I realize now that if one had an eye to see it, that little backyard world of mine held all the beauty and wonder of the entire universe. And I had all the time of a child to discover many secrets which I made my own and often later shared with astonished parents.

Mama enjoyed and encouraged my discoveries, but it was Papa who sent me to the encyclopedia and other books. Nature is great in books, but book knowledge cannot beat the pungent smell of a skunk, the humorous feat of a chameleon blowing up an orange-bubbled throat, the scolding of a blue jay, the feel of a wet frog or the soft fur of a bunny rabbit. The hound dog in the encyclopedia cannot look into your eyes and let you know he is all yours.

Just as I spend almost all my time now out of doors taking care of the some fifty-five miniature horses that support this Poor Clare monastery, I also spent the greater part of my childhood out of doors. Smells of nature were exciting to me. If I could not have seen some things, I would have known them just by their smells—our lilac bushes, honeysuckle vines by the front porch, the crushed pine needles whose bitterness I often tasted as well. Did you know a daddy longleg spider has a cloying sweet smell? I could detect a mouse nest before I ever saw it. Not one of Mama's rose bushes had the same perfume. Dahlias can make you sneeze if you inhale the blossom too deeply.

I knew what a pigeon was up to just by the inflection of his coo. And a turtle can grunt. I know. I had fifty of them when I was a little girl. Squirrels chatter and skunks whistle. Nobody told me that. An injured rabbit cries like a baby and frogs burp.

Not all rabbit hair feels the same. Snakes are dry, smooth, and the feel of their scales is intriguing. Rats' teeth are sharp and turtles haven't any. But never stick your finger into a turtle's open mouth. I have yet not to be surprised by a different blade of grass. I loved that broad-leafed grass-blade that stuck on your tongue like velcro if you rubbed it the wrong way.

Between Mama, Papa and Gus, the gardener, weeds had a very hard time surviving in my childhood backyard. I appreciated the beauty of our trees, shrubbery and flowers; but weeds did not escape my attention. They fascinated me as much as the cultivated flowers, yet I must admit that my folks had a way of allowing plant life to flourish as if only God did the work.

The exciting things that wiggled, crawled, hopped, or flew, the delicious earth odors, and the rustle and chirp of unhumans became a frequent means of drawing all things to myself as I sat quietly up high on a smooth limb of a pear tree whose bark still carries the wound of a jackknife where I carved a heart with the letters G-O-D and M-E.

He speaks to us in so many ways if only we would stop to listen to His silence as He makes His presence known. Gazing upward through the blossomed pear tree, I saw patches of blue and fleecy clouds and believed that He was up there somewhere, so they said. But somehow I sensed more than that. I knew He was all around me, saturating me with His odors, His sounds, His touch, His love—and I drowned myself in His beauty through my senses.

But the Kingdom of God is also within you, I came later in my spiritual growth to learn and appreciate. I will always know, however, that it was through His handiwork that I first found Him and it is through His handiwork that He so deeply continues to speak to me now.

I can from today's perspective also see the joining of the lines of another kind of circle—the training given my sister, brother, and me by our parents, training which prepared me for the discipline and the work which God had ahead for me.

From our earliest days we children were geared to strive for Christian perfection. But, believe me, I have no early claim to sainthood. Of the three children in the family, I gave my parents the most work. God decided to endow me, as a child, with blond curls and a slender build. (Oh, how all of that has changed.) But then He gave me this great love of nature and the outdoors, so what should He expect but a tomboy? And along with these qualities came all the trouble that little boys can get into.

Mama often said, "There is nothing you cannot do if you put your mind to it." I believed her and tried anything. Papa would say, "Use the brain God gave you." And I certainly came up with all sorts of ideas. Looking back, I can see that my application of my parents' remarks was often slightly stretched as a convenient license to further my own bright ideas. An inquisitive nature and a desire to investigate everything often led to unmitigated disaster.

Oh, I was justly punished for the ill effects of refutable conduct. I never remember resenting punishment meted out and always was made to remember it was for my good and that I deserved it.

"Blood runs thicker than water," Mama said. Then in her next breath, she would add, "No one will ever care for you the same as members of your own family." These oft-repeated statements usually followed the meting out of just punishment for some ungrateful act.

My brother, being close to me in age, was my almost constant companion. He was somewhat bashful and always good except when I got him into trouble. He was fun to be with and acquiesced to most of my ideas. He was—and is—very bright and always made me very proud of him. Sis, seven years older than I, was someone I looked up to with respect. She always seemed so grown up and Mama always trusted her decisions. She looked after Bruds and me.

Today my sister and brother remain central in my life. Each of them has been supportive and loving throughout all the challenges and the joys of my religious life. As a happy family unit growing up we were still certainly diverse individuals who managed "to do their own things," yet ultimately contributing to a harmonious and loving relationship which over the years has never lost the true meaning of family.

In spite of the constant concern we must have here at the Poor Clare Monastery to earn our daily livelihood and pay our debts while at the same time never veering from the daily purpose for our being—to remember that our very occupation is contemplative prayer—we find happiness everywhere. Franciscans are noted for joy. St. Francis would not stand for sadness in a person. Sadness was equated with sinfulness; only sin should make us sad.

Franciscans also believe in individuality. We pride ourselves here at the monastery on the fact that no two Franciscans are ever alike. That may seem contradictory at first glance of a photo of monks or sisters, all wearing the same kind of habits and cords. But look more closely. Some cords are on the beltline, some below it, and some above. Sizes and shape even make the common habit have a different look.

But truer still is the interior of the person. We have been encouraged to be true to ourselves, to be ourselves. That will happen if I accept myself the way God made me and others the way God made them. For instance, during my entire religious life, my spontaneity has never been squelched. In rounding off my rough edges and polishing me up to become a religious, God seems never to have allowed any of my originality to suffer. On the contrary, my fun-loving spirit, the playfulness of my childhood, and my innate sense of humor continues to haunt me and to get me into all the predicaments of old.

I remember on my first vacation home as a sister—I was nineteen years old—I wondered how I was supposed to act. I think it took a while for me to get used to myself as a professed sister. After some time I realized that one does not *act* like a sister. One either is or isn't a sister. I discovered that I was just another unique contribution to what makes up the sisterhood. Thence came the responsibility to act with the dignity of my calling and another glorious discovery that I am, and always will be, my same human self. The habit did not include a pair of wings nor a halo, nor had the world changed in my eyes. The same human battles I fought before I became a sister still exist. But now, with God's help, I am better equipped to handle situations—and, above all, to handle myself. Religious life has made me more true to myself, more true to others, and more true to the God to whom I belong.

I never thought a pair of cowboy boots on a nun could cause such excitement. But it seems that every bluebonnet tour that comes through Brenham wants the Monastery Miniature Horse Ranch on its agenda. We are now booked up a full year in advance. It is not uncommon for as many as 21,000 people to visit our monastery in any one year.

We also now have two horse shows per year. And no one is surprised when as many as eight hundred cars show up for a single horse show. CBS came for one show, and we made "CBS Evening News with Dan Rather." He evidently was amused seeing me driving a ladies' antique carriage drawn by a tiny buckskin miniature horse. I think that was the first time I ever saw him smile at the end of his television show.

We maintain an average herd of sixty horses at all times. Foals are born in the spring. Although we did lose that little foal early on, we have had great success with the baby horses since that time. We keep four or five pastures containing a group of mares matched with a stallion of our choice. The stallions measure 26, 28, or 29 inches. The mares range from 26 to 34 inches, most being 32. Our breeding stock have all been shown at one time or another. In fact, the ranch office is loaded with ribbons and trophies.

And we certainly have had some interesting times at the miniature horse shows. There was the time Sister Angela and I had gone to the national show in Kentucky. We were sitting up in the bleachers when the lady next to me exclaimed, "My, you sisters, since Vatican II, have become human." I had the whole grandstand in an uproar when I quickly turned to her and retorted in a loud voice, "Yes, ma'am, you know I was *born* that way." I had to laugh myself, for I hadn't expected it to be all that funny.

Another time we had just finished the national show held at the Texas State Fair in Dallas. Our stalls were well decorated with Monastery Miniature Horse Ranch signs as were our trailers and pickup truck. This sweet little lady came up to me and said, "Congratulations!" "Thank you," I responded, "but for what?" I was well aware that our horses had won many ribbons and trophies but I thought she might have had something particular in mind. "Oh," she exclaimed, "why for that clever presentation. The monastery theme. Why, you are even dressed the part in your nun's outfit." I hated to have to set the lady straight, but I had to tell her this wasn't a costume—we were for real.

There are, of course, sisters in the Catholic Church who do many active things: teach school, do hospital work, provide social services. We Poor Clares, however, are contemplative nuns, and we do not get paid for our praying. That is why we must support ourselves in whatever way is appropriate and that does not interfere with our prayer life. I have come to see that the contact which a few of the sisters have with the public, as a result of our raising and selling the little horses, is another way to share our treasures of the Spirit with others. God has given us this beautiful land upon which we live. It has astonished us as day by day we welcome more and more visitors who come from many parts of the country and from afar. Our own happiness is increased as so many people recognize the peace, the quietness, the joy of our simple way of life.

It was Reverend Bruce Nieli who spoke about a contemplative life like ours in words that help explain the contribution we want to make to the world.

> "It might be difficult," *he wrote,* "for many today to understand contemplation and the contemplative life. In our very pragmatic, secular, activity-oriented, worldly society, dedicating one's life exclusively to prayer in the celibate state seems odd indeed. Good! Alleluia! That's exactly what the contemplative life is supposed to do. It is meant to be countercultural. It is designed to challenge and confront all of us. Who in this contemporary scene does not need to be reminded that we live in a passing world, that our hearts should be set on a Kingdom that is not here, that if our hopes are set on this life only, we are the most pitiable of creatures. Our relatives, neighbors, friends, co-workers, fellow parishioners, and we ourselves, all run around leading unfocused lives looking for answers from today's computer which will be out-of-date tomorrow.
>
> Poor Clare Sisters call us all to contemplation. Whether we are married or single, young or old, introverted or extroverted, women or men, each of us has within our soul a quiet space, a Sacred Heart, an inner Mount Sinai, a cathedral where God is waiting for us to rest with Him. Most of us are unaware of this because our lives are filled with so much noise, both within and without. Rather may we seek first His kingdom. If our shepherd is the Lord, we shall never want."

As long as I can remember, I have wanted to belong to God. I have never felt otherwise. But, having discovered that God dwells within me, I desire nothing other than having Him use me for His greatest honor and glory. I believe He has done just that. And in so doing He has left me an ordinary person with my own peculiar characteristics. His abiding Presence never lessens His precious gift of humanism.

Belonging to God alone has enabled me to find Him in everyone and everything. To recognize Him in all things is such a joy and a feeling of completeness. I can describe it only haltingly in verse.

Tell A Vision

Inspired by early morning's
Rising Son
Against a grain-towered
Tree-lined sky
Where numbered days make
Ready, ways I sing a song
Before unsung, unknown.

Where once again my life
"Exempt from public haunt,
Finds tongues in trees,
Books in running brooks,
Sermons in stones,
And good in everything."*
O Brenham hills, by name
I name you now.

For if I do not tell a vision—
Whose tongue can tell?
I am His eyes, the eyes
That grasp the earthen beauty
Of the Son Who rose for me.
Duty, mine, to tell the blessed vision
To the tenant Who awaits, awaits.

You wait, O Risen Son
To glimpse the beauty
Now through Your other eyes.
Control the retina of my mind
To gather only Your reflections
So that deep within Your recess
All the beauty mine, Your beauty find.

*With apologies to Shakespeare

———————

Whatever we have accomplished upon this beautiful land has been through the mercy and bounty of the God whom we serve. "We are only earthen vessels which hold this treasure, to make it clear that such an overwhelming power comes from God and not from us" (from 2 Corinthians).

I shall never forget that traumatic moment when Madre Clara and I were advised: "Send the Cuban sisters two by two to other Poor Clare foundations." We had left that meeting in Minneapolis determined to establish a monastery for the refugee Cuban sisters. And today we can see that God has blessed us beyond our wildest expectations. We have a great future here as a contemplative community. Since we have been in Brenham, four wonderful women have joined us in this way of life. The proximity to the larger cities in Texas brings many visitors each day. This is all such a far cry from our first attempt so long ago to be included in the federation.

There are different kinds of growing up, the stages of physical growth being, of course, the most obvious. But more subtle than physical growth is the growth of wisdom, the growth of grace. There is the challenge to our consciousness.

God presents us with ordinary things and ordinary means for our growing up. Looking back now, it seems that nothing happened in my life by pure chance. All the incidents in my life contributed to my growing in wisdom and grace, in my becoming whole.

When I was a child, nothing escaped Mama's and Papa's attention, and a chance to practice virtue was often pointed out to us children, both by their words and example, but always in such a way to make us want to do the more rather than the less. And all these little golden opportunities came through ordinary daily occurences. It was a way of life, this growing in grace.

From this vantage point in my life, I can see how much the ordinary affairs of family living prepared me for the life God had planned for me. Both parents showed much interest in everything we did. We made our own entertainment. There was a togetherness that made us feel secure. We were proud of each other's accomplishments and were encouraged not to be wall flowers. Mama sang, and Sis accompanied her on the piano and recited poetry beautifully. Papa sang and was an accomplished orator. Bruds was an expert imitator, bird calls and pantomimes being his specialty. I played the piano.

Both parents were discriminating and made a point of getting to know the friends who came to play with us. I remember Papa as being very kind and gentle but very, very stern. He frowned a lot, especially when he did not approve of something we said or did. He was very proper, upright and unbendably righteous. He was brilliant and highly respected by us and many other people. Try as he might, he was never able to get the knack of fixing things around the home that required mechanical work. He was a good man. I never thought either of my parents could do anything wrong. The only defect I can ever remember was an occasional argument they would get into, usually concerning politics or a game of cards which Papa always took much too seriously. Sometimes Papa would get up and quietly leave the room if the ladies started gossip or talked about something of which he did not approve.

Mama had a very strong character. She was a decision maker, a planner, extremely generous, and a devoted mother who very rarely spent money on herself. She had a most beautiful singing voice. She loved the finer things and, although never going beyond a high school education, she could hold her own in any company.

As a happy family unit, we were still certainly diverse individuals who managed to do our own thing, ultimately contributing to a harmonious and loving relationship which over the years has never lost the true meaning of family.

———————————

It has been this sense of family that I have attempted to foster in the years I have been in the monastery. The talents of each sister are God-given. It is our duty to develop these. Not only is it our duty, but it is fun—fun to be shared. And the sharing increases the fun. There are certain things some people can do that others cannot do. It's smart to admit it and let it go at that. Try standing on your head if you do not believe me.

In the monastery we develop a certain awareness of God's presence without actually defining it as such. I learned early: "Where is God? God is everywhere." We believe this and actually act as if we do. Grace sharpens our awareness and thus the conscience becomes more sensitive.

All this growing in grace and wisdom is fed continually by the ordinary everyday things which come our way. Christ gave us this example by living thirty years of his thirty-three in an ordinary way, long before ever performing a public miracle. "He went down with them then, and came to Nazareth and was obedient to them. His mother meanwhile kept all these things in memory. Jesus, for his part, progressed steadily in wisdom and age and grace before God and men."

All the little golden opportunities to gain wisdom come through ordinary daily occurences in the monastery and on the grounds. It is a way of life, this growing in grace. Will we one day discover, perhaps, that saints are just ordinary people who do ordinary things extraordinarily well?

For me, the climax and stamp of approval upon our operations here in Brenham came with the letter I received one day most unexpectedly. I had driven the golf cart out to the rural mailbox with my Dalmatian, Apache, sitting beside me on the front seat. I think he enjoys seeing me reach into that mailbox. "You're going to have to sit here for a minute," I told Apache, "while I look through this international Poor Clare bulletin that has just come in the mail."

As I hurriedly scanned the bulletin which had just arrived, I was totally surprised at what I saw. A paragraph about our foundation.

"Apache, listen to this," I said excitedly to my attentive companion. "Listen to these words about our own monastery here in Brenham."

Reverend Dario Pili from the Franciscan Curia in Rome, who had recently visited all the Poor Clare monasteries in the United States, had written:

> The community of Brenham is extraordinary. Several years ago Poor Clares who had escaped from Cuba found refuge here. No kind of wall or tower could ever give the sense of solitude and silence, the sense of the infinite and the "distance" from the "world" which one breathes in Brenham. The miniature horses which live on the grounds give the place the biblical fascination of a deep relationship between human beings and animals, between humanity and the rest of creation, which was normal before the ecological disaster of original sin.

Sensing my exuberance upon reading these words, Apache gave a joyful leap off the golf cart and ran ahead of me as I drove back up to the monastery, one hand on the wheel, the other waving the paper on high. The wind whipped my singing into the trees: "Thank you, Father Dario, for this beautiful tribute!"

I slammed on the brakes and sat stark still. Even Apache stood at attention. The sun had suddenly broken through the trees and engulfed me with its warmth. Then came the realization—my whole life had been spent *challenged by God*. At that moment I cried out, "How I wish I could mount the roof of the monastery and shout out to all those who have yet not found Him: 'Taste and see that the Lord is Good! The Kingdom of God is within you.'"

The 13,000-square-foot Monastery of St. Clare building was designed and built by Charles Keese of Brenham in 1986. The layout of the monastery rooms is based on the same designs as the early European monasteries—the chapel, cell wings, and refectory are each separated for privacy. The beautiful chapel, which is at times opened for the public, becomes for the Poor Clare nun the center around which her whole life revolves.

We pose in our new chapel in the monastery at Brenham. Seated, left to right: Sisters Martha, Ascencion, Margaret Mary, Theresa Clare, Magdalena, Teresita, Celina. Standing, left to right: Sisters Rafael, Mary Joseph, Madre Clara, Mother Mary Bernadette, Sisters Holy Spirit, Angela, and Veronica.

The stained-glass windows were designed by the sisters and executed by Olszewski Stained Glass Studio of Corpus Christi. They are one-inch-thick faceted glass, chipped on the inside to give a shading. The theme of the windows was suggested by St. Francis' *Canticle of the Sun*, a copy of which is framed on the back wall of the chapel. This is a prayer composed by St. Francis in which he praises God through all the elements: brother sun, sister moon, the water, animals, stars, fire, flowers, and birds of the air.

I hold one of our first little miniature horses born in Brenham.

This is the latest picture of Sis, Bruds, and me. (Photo by Foschi Studio)

The garden in Brenham keeps us very busy. Madre and I share a hot day in the yard.

I water the plants while my habit and hat blow in the breeze.

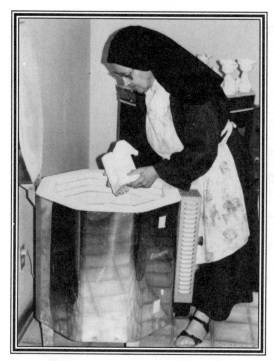

Sister Holy Spirit enjoys working in the ceramic studio. She paints the bisque figures and fires them in one of our three kilns in order to keep the gift shop shelves filled with attractive items.

All of us who are able take a hand at running the big tractor. Mowing the lawn is not too hard and really a lot of fun. There is more than one way to be a "flying nun."

Crowds love our minia-
ture horses at the horse
shows. (Photo by John
Herndon)

Although we do
have a dryer, we
prefer to give the
laundry a good sun-
bleaching when we
can. "Brother Sun"
does a good job,
helped by Sisters
Celina and Holy
Spirit.

The chapel becomes for the Poor Clare Nun the center in which her whole life revolves.

A priest celebrates Mass here every morning. At this time altar bread is consecrated and those present receive the Body of Christ in Communion. Extra breads are consecrated and reserved in the golden tabernacle on the altar. The sisters take turns every morning in adoration before this Blessed Sacrament.

As one enters the chapel the large crucifix over the altar dominates the front wall. This crucifix was carried onto the ferry by the Poor Clare Nuns who evacuated Cuba in 1960.

Here I am playing the piano again. I enjoy the use of musical instruments to enhance the liturgy.

Aside from their times of personal prayer, the sisters gather together in the chapel five times during the day to pray the Psalms, meditate, and say the rosary. Each sister has a designated place to kneel and keep her prayer books. There is also a place for visitors to join us in prayer.

A familiar sight to visitors, me with my protector, Apache Prince.
(Photo by Ken Appelt)

ORDER COUPON

CENTERPOINT PRESS
7I6 EUCLID STREET
HOUSTON, TEXAS 77009

Please send me _____ copies of *Sister Bernadette: Cowboy Nun from Texas*

Name_____

Address_____

City_____ State_____

ZIP_____

_____ books/$14.95 = _____

*shipping = _____

(Texas residents only add $1.27 per book) sales tax = _____

TOTAL: = _____

*Shipping
Include: $2.50 for 1st book
$1.50 for each additional book